MAGIC SQUARES

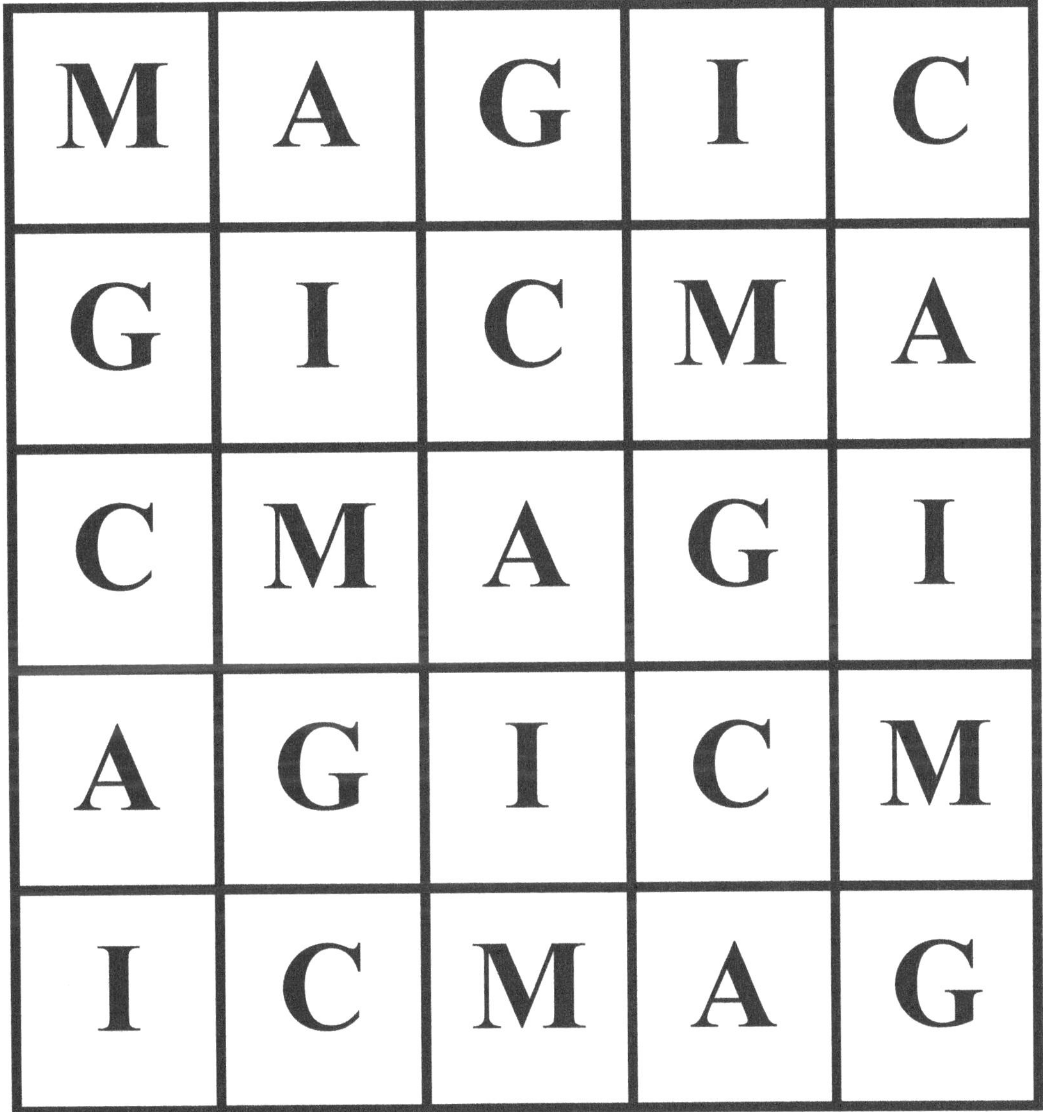

by

MARK S. FARRAR

To order additional copies, please contact us.
BookSurge, LLC
www.booksurge.com
1-866-308-6235
orders@booksurge.com

CONTENTS

Chapter		Page

APPENDICES

Acknowledgements are due to the following people:

- ➢ **Joe Riding**, who was my magical mentor and who first showed me the wonders of Magic Squares

- ➢ **Richard Stupple**, who, through a chance meeting, rekindled my dormant interest in magic and without whom this book would probably not have been written

- ➢ the other members of the **Northamptonshire Magicians' Club** who kept that rekindled spark alight

- ➢ **Pip Ayris**, who awakened my dormant interest in origami

- ➢ **David Berglas, Eddie Chapman, Mick Gurr, Alan Shaxon** and **Stephen Tucker**, who all contributed directly to this book

- ➢ all the other authors and magicians whose works I have cross-referenced or drawn upon

and, last but definitely not least, to my darling wife, **Rae**, without whose help and support this book would not have made it this far, and to whom this book is dedicated.

1.   Introduction

> *"I have often admired the mystical way of Pythagoras, and the secret magic of numbers"*
>
> *(Sir Thomas Browne, 1605 - 1682)*

The above quotation accurately describes my own fascination with numbers generally, but also with magic squares. I hope to show you in this book some of the many aspects of magic squares, and maybe kindle that spark in you too.

This book is a compilation and cross-reference of various magic square literature. Its focus is on a detailed, non-mathematical analysis of magic squares and the different varieties and presentations of magic squares. It contains:

-       a few general comments about all magic squares

-       some general formulae

-       a detailed analysis of the most common orders of magic square (i.e. 3 x 3, 4 x 4 and 5 x 5)

-       notes on some of the variations on the magic square theme

-       several presentations and routines for magic squares, including both references to other people's routines and some ideas of my own

-       a list of books that I have referenced within this booklet.

All of this is supported by Appendices containing some of the most detailed analyses of magic squares that have probably ever appeared in print. Four of these - Appendices A, B, C and D - have been created using the analytical powers of modern computers.

2.　　 History

The first question has to be, "What **is** a magic square?" Essentially, there are two types.

The early ones included a series of letters, arranged in a square, to spell certain words. One of the most well-known of this variety, throughout the Western world at any rate, is:

S	A	T	O	R
A	R	E	P	O
T	E	N	E	T
O	P	E	R	A
R	O	T	A	S

Note that the five Latin words appear in the same order both horizontally and vertically, and that the five rows may also be read palindromically. They also form a sentence - "Arepo, the sower, guides the wheels at work". A copy of this square has been found in the ruins of Pompeii, dating from the first century A.D, and also in a Roman villa in England.

It also appears, together with a very similar square, in "The Necromantic Grimoire Of Augustus Rupp" by Anthony Raven (see Chapter 10, References, for further information).

Such squares were originally used as religious symbols, as special, magical properties were ascribed to them.

For instance, the letters in the above magic square can be rearranged to spell the word Paternoster, meaning Our Father, both horizontally and vertically, with the letters A and O left over, these letters representing the Alpha and the Omega, which is another term for God:

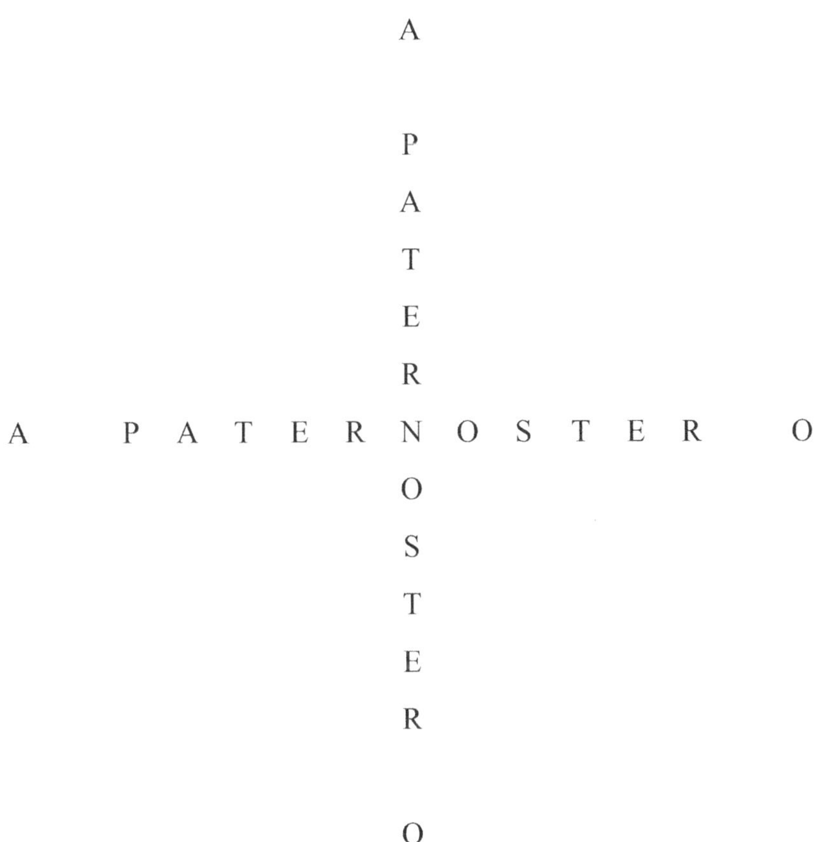

These gave way to magic squares that consist of a series of numbers arranged in a square in such a manner that the sum of each row, each column and of both the corner diagonals adds up to the same amount, which is called the "magic total".

Various other features **may** be added to such a square which may enhance its mathematical interest, but these features are considered non-essential.

Magic squares have fascinated mankind throughout the ages, with examples being found in:

-       Chinese literature dating from as early as 2800 B.C., when a magic square known as the "*Loh-Shu*", or "scroll of the river Loh", was invented by Fuh-Hi, the mythical founder of Chinese civilisation:

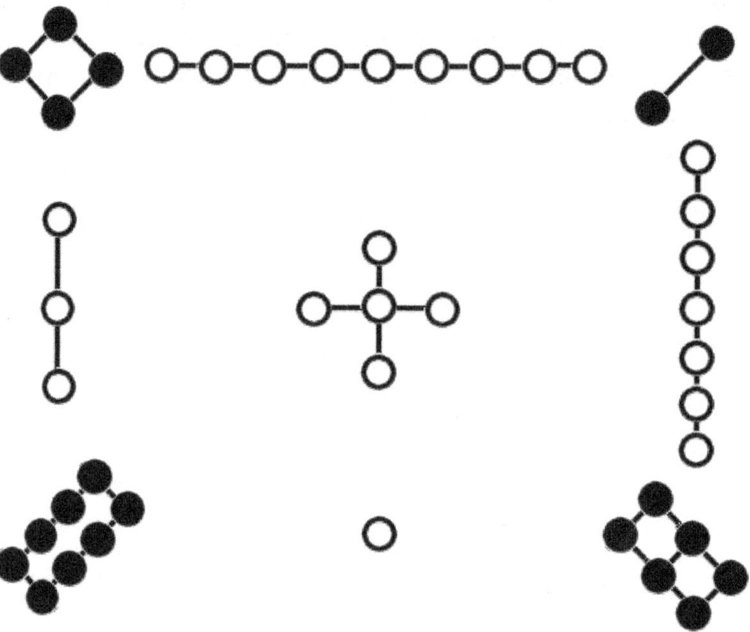

Note that:

-      the odd numbers are shown as white dots, representing "yang" symbols (i.e. the emblem of heaven)

-      the even numbers are shown as black dots, representing "yin" symbols (i.e. the emblem of earth).

It is interesting that the idea of odd numbers belonging to heaven and even numbers belonging to earth was also later shared by the ancient Greeks, even though magic squares were developed entirely independently by these two civilisations.

-      Greek writings dating from about 1300 B.C.

-      the works of Theon of Smyrna in 130 A.D.

-      use by Arabian astrologers in the ninth century when drawing up horoscopes

-      Arabic literature, written by Abraham ben Ezra, dating from the eleventh century

- India, dating from the eleventh or twelfth century, where the earliest fourth order magic square was found, in Khajuraho:

7	12	1	14
2	13	8	11
16	3	10	5
9	6	15	4

- the writings of the Greek mathematician, Emanuel Moschopulus, whose works now reside in the National Library in Paris

- more recently, magic squares appeared in Chinese literature during the latter part of the posterior Chou dynasty (951 - 1126 A.D.) or the beginning of the Southern Sung dynasty (1127 - 1333 A.D.).

- the works of Cornelius Agrippa, a German physician and theologian from the sixteenth century, who constructed seven magic squares, of orders three to nine inclusive, which he associated with the seven planets then known (including both the Sun and the Moon)

- art, with the relatively well-known magic square which can be found in Albert Dürer's engraving "Melencolia" and is shown in Figure 1 on Page 6.

  Note that the middle two cells on the bottom row indicate the year in which this work was made, i.e. 1514 A.D.

- a detailed French work, published in 1838 A.D.

They were also frequently found in various cultures, for example, Egypt and India, engraved on stone or metal and worn as talismans, the belief being that magic squares had astrological and divinatory qualities.

Figure 1 - "*Melencolia*" by Albert Dürer

3.   General Information

1.   Types

There are two general classes of magic square, namely, "**odd**" squares and "**even**" squares:

-     by "odd" squares, it is meant that there is an odd number of cells on each side of the magic square (e.g. 3 x 3 and 5 x 5 magic squares)

-     in a similar manner, an "even" square is one that has an even number of cells on each side of the magic square (e.g. 4 x 4 and 6 x 6 magic squares).

      "Even" magic squares may be further sub-divided:

      -     "singly even" magic squares are ones where the number of cells on each side of the magic square is evenly divisible by two, but not by four (i.e. 6 x 6 and 10 x 10 magic squares)

      -     "doubly even" magic squares are ones where the number of cells on each side of the magic square is evenly divisible by both two and four (i.e. 4 x 4 and 8 x 8 magic squares).

For each order of magic square, there is always a "basic" magic square, which uses the numbers 1 to n^2, where "n" is the number of cells on each side of the magic square, also known as the order of the magic square. For example, a "basic" 3 x 3 magic square will use the numbers from 1 to 9 inclusive, whilst a "basic" 4 x 4 magic square will use the numbers from 1 to 16 inclusive.

2.   "Magic Totals"

This "basic" magic square will sum to a "magic total" which may be calculated by the formula $(n^3 + n) / 2$, where, again, "n" is the number of cells on each side of the magic square. For example, a "basic" 3 x 3 magic square will sum to 15, i.e. $(27 + 3) / 2$; similarly, a "basic" 4 x 4 magic square will sum to 34, i.e. $(64 + 4) / 2$.

These rules hold true for **all** "basic" magic squares, whether they are "odd" or "even".

However, for certain "basic", "odd" magic squares, the "magic total" may also be calculated as the value in the centre cell multiplied by "n", where "n" is the number of cells on each side of the magic square. This method may only be applied to "odd" magic squares that have been created according to certain methods, such as the De La Loubère method (see Chapter 5, An Analysis Of 3 x 3 Magic Squares).

3.　　**Further Classifications**

Magic squares may be further classified as:

- "semi-magic", which means that the magic square is not a full, true magic square; this can occur, for example, where the rows and columns sum to the "magic total", but the corner diagonals do not

- "associative" (or "regular"), which means that the sum of squares that are diametrically equidistant from the square's central point add up to the sum of the first and last numbers used in completing the square

- "panmagic" (or "diabolic"), which means that the "broken" diagonals also sum to the "magic total", as well as the two corner diagonals; see Chapter 8, Variations, for further details

- "bimagic", which means that each cell in a magic square may be replaced by the square of the cell's value and still remain a magic square.

4.   **General Formulae**

There are two main methods of creating a magic square that sums to a specified "magic total", but note that, depending on the method used to create the magic square, it may need to be "transposed" to make it easily usable for certain presentations (see Chapter 6, An Analysis Of 4 x 4 Magic Squares and Chapter 7, An Analysis Of 5 x 5 Magic Squares):

1.   **"Simple" Formula To Create A Magic Square Adding To A Specified "Magic Total"**

This "simple" formula is the easier, in that less mental arithmetic is required. Details of this formula may be found, for example, in "The Magic Book" by Harry Lorayne.

The following three steps should be followed:

1.   The first $(n^2 - n)$ numbers in the magic square may be written in straight away, where "n" is the number of cells on each side of the magic square, as per the "basic" magic square.

2.   The next number is derived by subtracting the result of the difference between $((n^2 - n) + 1)$ and the "basic" magic square's "magic total", from the required "magic total".

3.   The remaining numbers are successively increased by one, starting from the number derived in Step 2.

As an example, suppose that a 4 x 4 magic square is asked for with a required total of 51.

The result of Step 1 would be that the numbers 1 to 12 (i.e. 16 - 4) are written into their normal cells.

The result of Step 2 would be that 30 is written into the square usually containing 13 (i.e. the difference between 34, the "basic" 4 x 4 magic square's "magic total" and 13, subtracted from 51).

Finally, the result of Step 3 would be that the numbers 31, 32 and 33 are written into the squares usually containing 14, 15 and 16.

The resulting magic square could be, depending on the method used to create the magic square:

9	6	3	33
4	32	10	5
31	1	8	11
7	12	30	2

Note that numbers less than the "basic" magic square's "magic total" **may** be chosen as the required total, but this will result in either numbers being repeated and/or some numbers being zero or negative. For example, imagine that a 5 x 5 magic square is requested that adds up to 42. This could be the result, depending on the method used to create the magic square:

0	6	19	2	15
4	12	2	8	16
10	18	1	14	-1
11	1	7	20	3
17	5	13	-2	9

As you can see, this magic square contains negative numbers, a zero, and some repeated numbers. However, it still sums to the requested total of 42.

Note also that magic squares that sum to a total greater than their "basic" magic square's "magic total" will never be "regular".

The advantage of this "simple" formula is that numbers may be written into the magic square as soon as a spectator calls out the required "magic total".

The disadvantages are that:

1. The greater the required "magic total", the bigger the gap between the "normal" numbers and the highest "n" "adjusted" (or "key") numbers, where "n" is the number of cells on each side of the magic square. This makes the magic square look "unbalanced".

2. There are fewer combinations of cells that sum to the "magic total" than exist using the more complicated formula (detailed below). Further details may be found in the next three Chapters.

2. **"Complex" Formula To Create A Magic Square Adding To A Specified "Magic Total"**

This is a slightly more complicated method, which uses the following formula, details of which may be found, for example, in "The Amazing Magic Square And Master Memory Demonstration" by Orville Meyer:

1. First subtract the "basic" magic square's "magic total" from the required total.

2. Divide the result of Step 1 by "n", where "n" is the number of cells on each side of the magic square.

3. Add the integer part (i.e. ignore any remainder) of the result of Step 2 to **all** cells in the "basic" magic square.

4. **In addition**, add the remainder (i.e. ignore the integers before the decimal point) of the result from Step 2 to the "n" cells with the **highest** numbers.

As an example, suppose that a 4 x 4 magic square is asked for with a required total of 51. The result of Step 1 would be 17 (i.e. 51 - 34). The result of Step 2 would be 4 remainder 1.

The result of Step 3, therefore, would be that 4 (i.e. the integer part of the result of 17 divided by 4) is added to **all 16** cells in the "basic" magic square.

Step 4 would result in an additional 1 being added to the cells in the "basic" magic square that contain 13, 14, 15 and 16; this is because the result of Step 2 gives a value of 4, remainder 1.

The resulting magic square would therefore be, depending on the method used to create the magic square:

13	10	7	21
8	20	14	9
19	5	12	15
11	16	18	6

Note that numbers less than the "basic" magic square's "magic total" **may** be chosen as the required total, but this will result in either numbers being repeated and/or some numbers being zero or negative. For example, imagine that a 5 x 5 magic square is requested that adds up to 47. This could be the result, depending on the method used to create the magic square:

17	3	16	-1	12
1	9	19	5	13
7	15	-2	11	16
8	18	4	17	0
14	2	10	15	6

As you can see, this magic square contains negative numbers, a zero, and some repeated numbers. However, it still sums to the requested total of 47.

Note also that magic squares that sum to a total greater than their "basic" magic square's "magic total" will only be "regular" if their "magic total" is evenly divisible by "n", where "n" is the number of cells on one side of the magic square, after subtracting the "basic" magic square's "magic total" from the required total. This is also the only situation where the numbers used in constructing the magic square will be consecutive.

The disadvantage of this "complex" formula is that numbers may not be written into the magic square as soon as a spectator calls out the required "magic total", since, potentially, every single cell will contain an "adjusted" number.

The advantages are that:

1.   The magic square will always look "balanced", since there will never be a large gap between any of the numbers. The gap will never be greater than "n", where "n" is the number of cells on each side of the magic square.

2.   There are potentially more combinations of cells that sum to the "magic total" than exist using the simpler formula (detailed above). Further details may be found in the next three Chapters.

3.   **Formula To Calculate The "Magic Total" Given The Starting Number**

Provided that **consecutive** numbers are used, the general calculation is:

1.   Multiply the lowest number used (i.e. the one given by the spectator) by "n", where "n" is the number of cells on each side of the magic square to be created.

2.   Add the "magic total" for the "basic" magic square of that order.

3.   Subtract "n".

For example, if a 4 x 4 magic square is to be constructed starting with the number 23, the predicted total would be (23 x 4) + 34 - 4, which results in a final figure of 122.

Note that, where the lowest number used is one, then this formula gives the same result as the formula given in Chapter 3, General Information, above.

4.   **"Disguised" Magic Squares**

All magic squares may be rotated and/or reflected to create what appears to be a new magic square. However, to use the terminology of Jim Moran in his book "The Wonders Of Magic Squares", such rotations and reflections are "merely the original magic square in disguise".

When counting the number of different magic squares, such "disguised" magic squares are usually ignored. However, trying to detect that one magic square is simply another one in disguise is not particularly easy when looking at the numbers that make up that magic square.

A much easier method is to draw what Jim Moran calls a "sequence diagram", which is created by drawing a straight line from one number to the next, i.e. from 1 to 2, from 2 to 3, and so on. The final line would be drawn from the highest number back to 1.

For example, a "sequence diagram" for the magic square shown in Albert Dürer's "Melencolia" would look like this:

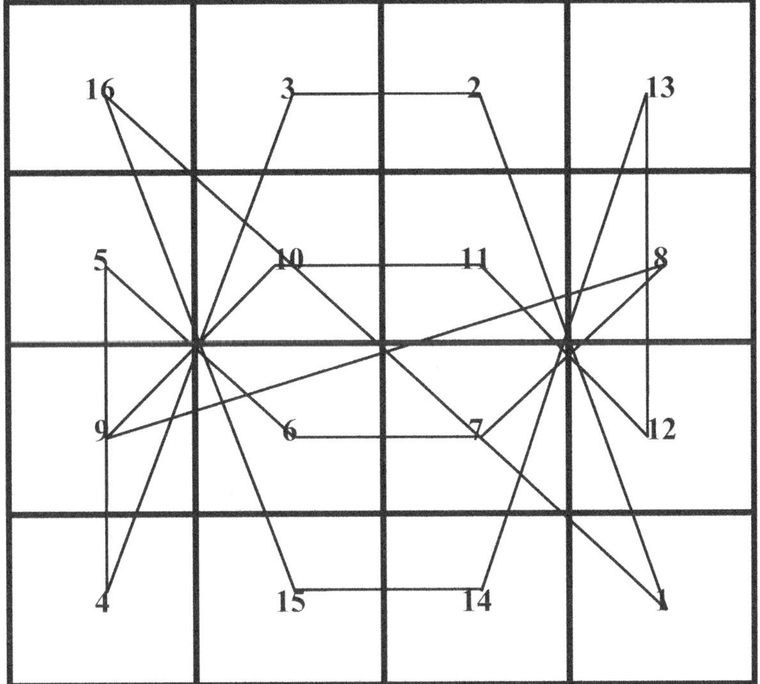

If this magic square is rotated through ninety degrees clockwise and a "sequence diagram" drawn for it:

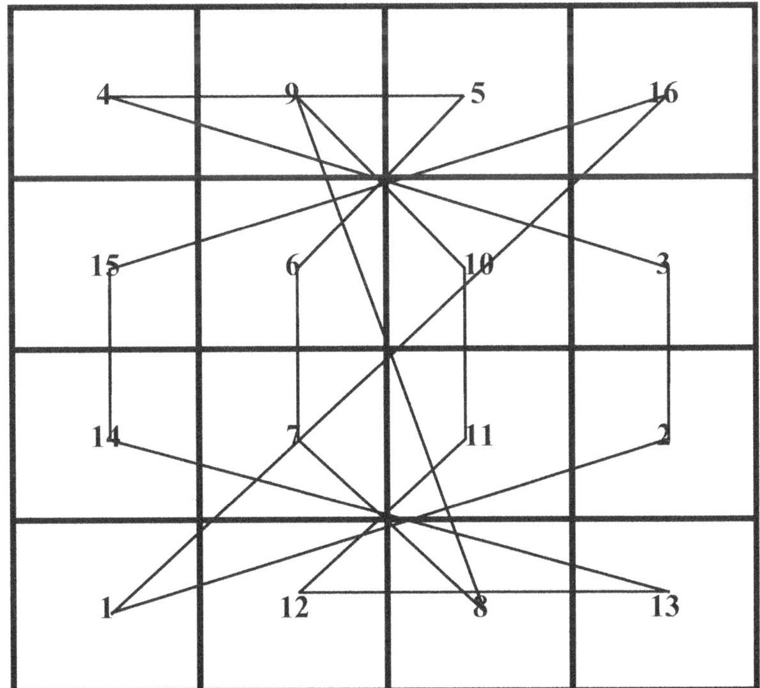

it is relatively easy to detect that it is really the same magic square.

More details may be found in Jim Moran's book, mentioned above.

5.   An Analysis Of 3 x 3 Magic Squares

With the exception of the **extremely** trivial 1 x 1 magic square, a 3 x 3 magic square is the smallest magic square that it is possible to form. The "basic" 3 x 3 magic square, using the numbers 1 - 9 inclusive, is as follows:

8	1	6
3	5	7
4	9	2

This magic square:

-    is "regular" (because each pair of cells that are diametrically equidistant from the centre cell add up to 10, which is the sum of the first and last numbers used)

-    is **not** "panmagic" (because the "broken" diagonals, i.e. the cells containing 1, 7 and 4, and 1, 3 and 2, do not add up to the magic square's "magic total").

Various mathematicians (e.g. Cornelius Agrippa, Poignard, Ozanam, Bachet de Mézeriac, Devedec, Philippe De La Hire, Professor Scheffler, Benjamin Franklin) throughout the years have devised methods of creating magic squares, but the simplest rules for forming such a "basic" magic square, sometimes called the "staircase" method and devised by De La Loubère in the late seventeenth century, are as follows:

1.   Write the number 1 in the centre cell on the top row.

2.   Move one cell up and one cell to the right. (To do this, you have to assume that the top row "wraps round" to the bottom row, and the right column "wraps round" to the left column.)

3.   If this cell is empty, write in the next highest number in the sequence.

4.   If this cell is not empty, move down one cell, within the same column, "wrapping round" from the bottom row to the top row if necessary.

5.   Repeat steps 2 to 4 until all cells have been completed. The largest number in the sequence should be in the middle of the bottom row; if this is not the case, then you have probably made a mistake somewhere.

In the above example of a "basic" 3 x 3 magic square, the above steps would result in the following:

1.   The number 1 would be written in the middle cell of the top row (as per Step 1).

2.   The number 2 would be written in the last cell of the bottom row (as per Steps 2 and 3, having "wrapped round" from the top row to the bottom row).

3.   The number 3 would be written in the first cell of the middle row (as per Steps 2 and 3, having "wrapped round" from the right column to the left column).

4.   The number 4 would be written in the first cell of the last row (as per Steps 2 and 4, since the cell one up and to the right of the cell which contains the number 3 already contains the number 1).

5.   The number 5 would be written in the middle cell of the middle row (as per Steps 2 and 3).

6.   The number 6 would be written in the last cell of the top row (as per Steps 2 and 3).

7.   The number 7 would be written in the last cell of the middle row (as per Steps 2 and 4, since the cell one up and to the right of the cell which contains the number 6 already contains the number 4, "wrapping round" from the top row to the bottom row and from the right column to the left column).

8.   The number 8 would be written in the first cell of the top row (as per Steps 2 and 3, having "wrapped round" from the right column to the left column).

9.   The number 9 would be written in the middle cell of the bottom row (as per Steps 2 and 3, having "wrapped round" from the top row to the bottom row).

My fascination with this generic method and its application to "odd" magic squares of any order, even large ones, led me to create the computer program contained in Appendix E. This program will create a "basic", "odd" magic square for any order, and it will run quite easily on most home or business computers.

Out of the 84 **possible** combinations of three squares from nine, it can easily be seen that **eight** of these add up to this magic square's "magic total" (i.e. 15): namely, each of the three rows, each of the three columns, and both of the two corner diagonals.

In fact, there are **only** eight ways in which the magic total of 15 can be obtained by adding three of the numbers 1 to 9; each of these eight ways occurs once in this magic square. There is, therefore, **only one** magic square of order three, excluding all rotations and reflections of the same square; this fact may be proved using elementary mathematics and logic (see "Mathematical Recreations" by Maurice Kraitchik or "In Code" by Sarah & David Flannery for details).

If the magic square sums to a "magic total" other than the "basic" magic square's "magic total":

1.   If the "complex" formula is used, and if the required total is **not** divisible by three (see Chapter 4, General Formulae, for details of how to create magic squares which sum to a number other than the "basic" magic square's "magic total"), then the number of combinations in the resulting magic square that add up to the required total will be **seven**, not eight. This is because one of the two corner diagonals (the one from the top right to the bottom left) does not contain any of the three "key" numbers (i.e. 7, 8 and 9), therefore this diagonal will sum to a value that is either one or two less than the required total. In this instance, therefore, a "semi-magic" magic square will be created.

2.   If the "simple" formula is used, every magic square other than the "basic" magic square will always sum to the "magic total" in exactly seven combinations. This is because the corner diagonal which runs from the top right to the bottom left, containing the numbers 4, 5 and 6, and therefore summing to 15, does not contain any "key" numbers. These will always be, therefore, "semi-magic" magic squares.

6. An Analysis Of 4 x 4 Magic Squares

A 4 x 4 magic square is the smallest "even" magic square that it is possible to form. The "basic" 4 x 4 magic square, using the numbers 1 - 16 inclusive, is as follows:

16	2	3	13
5	11	10	8
9	7	6	12
4	14	15	1

This magic square:

-    is "regular" (because each pair of cells that are diametrically equidistant from the centre cell add up to 17, which is the sum of the first and last numbers used)

-    is **not** "panmagic" (because the "broken" diagonals, e.g. the cells containing 15, 7, 5 and 13, or 14, 9, 8 and 3, do not add up to the square's "magic total").

The simplest rules for forming such a "basic" 4 x 4 magic square, devised by Cornelius Agrippa in the early sixteenth century, are as follows:

1.   Write the numbers 1 - 16 in the 16 cells in ascending numerical order, starting in the top left cell and finishing in the bottom right cell.

2.   The numbers that are not on one of the two corner diagonals (i.e. 2, 3, 5, 8, 9, 12, 14 and 15) are now in the correct positions.

3.   The remaining eight numbers (i.e. 1, 6, 11 and 16, and 4, 7, 10 and 13) should now be moved to new squares. Each of these numbers should be exchanged with their diametrically opposite number (i.e. the numbers 1 and 16 are exchanged, the numbers 4 and 13 are exchanged, the numbers 6 and 11 are exchanged, and the numbers 7 and 10 are exchanged).

A similar method may be used to create any "doubly-even" magic square.

According to "The Penguin Dictionary Of Curious And Interesting Numbers", by David Wells, there are **880** magic squares of order four, **excluding** all rotations and reflections of the same square. Of these 880, **48** are "panmagic".

Out of the 1,820 **possible** combinations of four squares from 16, it can be calculated that **86** of these add up to this magic square's "magic total" (i.e. 34). Like the "basic" 3 x 3 magic square, these include each row, each column, and both of the two corner diagonals. However, in addition to these 10 combinations, there are up to 76 other combinations that will sum to the "magic total".

If the magic square sums to a "magic total" other than the "basic" magic square's "magic total", the **exact** number will vary, depending on whether the "simple" or "complex" formula was used to create the magic square:

1.   If the "complex" formula was used, the **exact** number will vary, depending on the remainder after 34 has been subtracted from that "magic total" and the result of that subtraction has been divided by four (see Chapter 4, General Formulae, for details of how to create magic squares which sum to a number other than the "basic" magic square's "magic total"):

Remainder	Number Of Combinations
0	86
1	78
2	74
3	68

See Appendix A for a complete analysis of the different combinations of four squares that will add up to the chosen total. I believe that this is the first time that such a detailed analysis has been published; it has not been widely recognised amongst magicians that there are so **many** different combinations that will sum to the selected total. (Note that this Appendix assumes that each cell is labelled from "A" through to "P", from the top left cell down to the bottom right cell.)

Note that there are **60** combinations that are common to all 4 x 4 magic squares, regardless of the remainder. These combinations are highlighted on each page of Appendix A.

2.   If the "simple" formula was used, then a subset of the combinations obtained via the "complex" formula will sum to the "magic total". The

number of combinations will vary from 86 down to 60. See Appendix B for a summary of these combinations.

Note that there are also **60** combinations that are common to all 4 x 4 magic squares when the "simple" formula is used.

However, the following magic square is more adaptable for presentation purposes. This is the above magic square, which has been "transposed" according to specific rules (which may be found in "Mathemagic" by Royal Vale Heath or "Greater Magic" by John Northern Hilliard):

9	6	3	16
4	15	10	5
14	1	8	11
7	12	13	2

The effect of the transposition is that each row, each column and both of the two corner diagonals each contains one and only one of the four highest numbers (i.e. 13, 14, 15 and 16).

Note that this magic square is:

-       no longer "regular" (because each pair of cells that are diametrically equidistant from the centre cell do **not** add up to 17, which is the sum of the first and last numbers used)

-       **is** "panmagic" (because the "broken" diagonals, e.g. the cells containing 6, 10, 11 and 7, or 3, 15, 14 and 2, **do** add up to the square's "magic total").

7.　　An Analysis Of 5 x 5 Magic Squares

The "basic" 5 x 5 magic square, using the numbers 1 - 25 inclusive, is as follows:

17	24	1	8	15
23	5	7	14	16
4	6	13	20	22
10	12	19	21	3
11	18	25	2	9

Like the "basic" 3 x 3 magic square, this magic square **is** "regular", but it is **not** "panmagic".

The rules for forming this "basic" 5 x 5 magic square are the same as for creating a "basic" 3 x 3 magic square (see Chapter 5, An Analysis Of 3 x 3 Magic Squares, above). In fact, these rules may be used to create **any** size, or order, of "odd" magic square.

There is an alternative method of constructing a 5 x 5 magic square that allows the spectator to place the number of his choice (in the range 1 to 25) in the square of his choice. The rest of the square is then completed, using similar rules to those described in Chapter 5, An Analysis Of 3 x 3 Magic Squares, above. Again, to create a magic square using this method, you have to imagine that the top row "wraps" round to the bottom row and the right column "wraps" round to the left column. The rules are:

1.　　If the number last entered is evenly divisible by five, write the next number in the cell two squares to the right of the last number, "wrapping" round from the right column to the left column if necessary.

2.　　Otherwise, write the next number **two** cells up and one to the right, again, "wrapping" round as required.

The next number to be written after 25 is 1.

A sample magic square, where the first number written was in the middle cell of the second row (i.e. 11), using this method is shown below:

14	2	20	8	21
10	23	11	4	17
1	19	7	25	13
22	15	3	16	9
18	6	24	12	5

Further details may be found in "Mathematical Magic" by William Simon.

According to "The Penguin Dictionary Of Curious And Interesting Numbers", by David Wells, there are **275,305,224** magic squares of order five, **excluding** all rotations and reflections of the same square.

Out of the 53,130 **possible** combinations of five squares from 25, it can be calculated that **1,394** of these add up to this magic square's "magic total" (i.e. 65). Like the "basic" 3 x 3 and 4 x 4 magic squares, they include each row, each column, and both of the two corner diagonals. However, in addition to these 12 combinations, there are up to 1,382 other combinations that will sum to the "magic total".

If the magic square sums to a "magic total" other than the "basic" magic square's "magic total", the **exact** number will vary, depending on whether the "simple" or "complex" formula was used to create the magic square:

1. If the "complex formula was used, the **exact** number will vary, depending on the remainder after 65 has been subtracted from that "magic total" and the result of that subtraction has been divided by five (see Chapter 4,

General Formulae, for details of how to create magic squares which sum to a number other than the "basic" magic square's "magic total"):

Remainder	Number Of Combinations
0	1,394
1	1,342
2	1,291
3	1,243
4	1,198

See Appendix C for a complete analysis of the different combinations of five squares which will add up to the chosen total. I believe that this is the first time that such a detailed analysis has been published; it has not been widely recognised amongst magicians that there are so **many** different combinations that will sum to the selected total. (Note that this Appendix assumes that each cell is labelled from "A" through to "Y", from the top left cell down to the bottom right cell.)

Note that there are **873** combinations that are common to all 5 x 5 magic squares, regardless of the remainder. These combinations are highlighted on each page of Appendix C.

2.   If the "simple" formula was used, then a subset of the combinations obtained via the "complex" formula will sum to the "magic total". The number of combinations will vary from 1,394 down to 873. See Appendix D for a summary of these combinations.

Note that there are also **873** combinations that are common to all 5 x 5 magic squares when the "simple" formula is used.

The following magic square has also been transposed:

23	6	19	2	15
4	12	25	8	16
10	18	1	14	22
11	24	7	20	3
17	5	13	21	9

The effect of the transposition is that each row, each column and both of the two corner diagonals each contains one and only one of the five highest numbers (i.e. 21, 22, 23, 24 and 25).

Note that this magic square is:

-       no longer "regular" (because each pair of cells that are diametrically equidistant from the centre cell do **not** add up to 17, which is the sum of the first and last numbers used)

-       **is** "panmagic" (because the "broken" diagonals, e.g. the cells containing 6, 10, 11 and 7, or 3, 15, 14 and 2, **do** add up to the square's "magic total").

8.   Variations

1.   6 x 6 Magic Squares

The next order of magic square to be discussed is the 6 x 6 magic square, an example of which is:

35	1	6	26	19	24
3	32	7	21	23	25
31	9	2	22	27	20
8	28	33	17	10	15
30	5	34	12	14	16
4	36	29	13	18	11

Such a magic square, which is a "singly-even" magic square, does not conform to the usual rules for creating either "odd" or "even" magic squares.

Various methods exist for creating such a magic square, but the simplest, devised by Ralph Strachey, is to be found in Royal Vale Heath's "Mathemagic", although a more detailed analysis can be found in either:

-        "Greater Magic" by John Northern Hilliard

-        "Magic Squares And Cubes" by W.S. Andrews

-        "The Moscow Puzzles" by Boris Kordemsky.

For the above 6 x 6 magic square, this method involves:

1. Splitting the 36 squares into four 3 x 3 mini-squares.

2. Completing the top left mini-square using the numbers 1 to 9, as per an ordinary 3 x 3 square.

3. Completing the lower right mini-square using the numbers 10 to 18, as per an ordinary 3 x 3 square.

4. Completing the upper right mini-square using the numbers 19 to 27, as per an ordinary 3 x 3 square.

5. Completing the lower left mini-square using the numbers 28 to 36, as per an ordinary 3 x 3 square.

6. Finally, transposing the three pairs of numbers 8 and 35, 5 and 32, and 4 and 31.

There are 32,134 combinations of six cells in this "basic" 6 x 6 magic square that sum to the "magic total" of 111 - far too many to show in an appendix!

It is interesting to note, also, that if the contents of each of the 36 cells of a "basic" 6 x 6 magic square are summed, the total is 666, the so-called "Number of the Beast". A new effect based on this fact is shown in Chapter 9, Presentation/Routines, below.

2. **Bordered Magic Squares**

A bordered magic square is one that remains "magic" when the outer border of numbers is removed. For example, the following order five magic square, with a "magic total" of 65, is a bordered magic square:

21	1	2	22	19
23	16	9	14	3
6	11	13	15	20
8	12	17	10	18
7	25	24	4	5

becomes an order three magic square, with a "magic total" of 39, if the outer edge is removed:

16	9	14
11	13	15
12	17	10

Larger bordered magic squares may be created, where several outer borders may be removed, one at a time, each leaving a smaller magic square.

A full mathematical analysis of such magic squares may be found in:

-      "Mathematical Recreations" by Maurice Kraitchik

-      "New Recreations With Magic Squares" by William H. Benson & Oswald Jacoby.

A less technical description may be found in, for example, "Creative Puzzles Of The World" by Pieter Van Delft & Jack Botermans.

3.   **Magic Cubes & Other Shapes**

Magic cubes are effectively three-dimensional magic squares, which are clearly more difficult to construct since the relationships from one face of the cube to another are more complex and less easily adjusted in relation to each other in order to sum to the "magic total".

There is:

-      an entire chapter devoted to magic cubes in "Magic Squares And Cubes" by W.S. Andrews

-      an entire book, called "Magic Cubes", devoted to them by William H. Benson & Oswald Jacoby

-      an interesting presentation, called "Wonder Blocks", using 16 "magic cubes" in "Magical Fun With Magic Squares" by Dr. Frank Blaisdell

-      a routine called "The Magic Block" in Royal Vale Heath's "Mathemagic", using the magic cube on the next page that first appeared in Ripley's "Believe It Or Not" in The New York American in October 1932:

<table>
<tr><td></td><td></td><td></td><td></td><td>87</td><td>12</td><td>86</td><td>9</td><td></td><td></td><td></td><td></td></tr>
<tr><td></td><td></td><td></td><td></td><td>82</td><td>13</td><td>83</td><td>16</td><td></td><td></td><td></td><td></td></tr>
<tr><td></td><td></td><td></td><td></td><td>11</td><td>88</td><td>10</td><td>85</td><td></td><td></td><td></td><td></td></tr>
<tr><td></td><td></td><td></td><td></td><td>14</td><td>81</td><td>15</td><td>84</td><td></td><td></td><td></td><td></td></tr>
<tr><td>63</td><td>36</td><td>62</td><td>33</td><td>79</td><td>20</td><td>78</td><td>17</td><td>95</td><td>4</td><td>94</td><td>1</td></tr>
<tr><td>58</td><td>37</td><td>59</td><td>40</td><td>74</td><td>21</td><td>75</td><td>24</td><td>90</td><td>5</td><td>91</td><td>8</td></tr>
<tr><td>35</td><td>64</td><td>34</td><td>61</td><td>19</td><td>80</td><td>18</td><td>77</td><td>3</td><td>96</td><td>2</td><td>93</td></tr>
<tr><td>38</td><td>57</td><td>39</td><td>60</td><td>22</td><td>73</td><td>23</td><td>76</td><td>6</td><td>89</td><td>7</td><td>92</td></tr>
<tr><td></td><td></td><td></td><td></td><td>71</td><td>28</td><td>70</td><td>25</td><td></td><td></td><td></td><td></td></tr>
<tr><td></td><td></td><td></td><td></td><td>66</td><td>29</td><td>67</td><td>32</td><td></td><td></td><td></td><td></td></tr>
<tr><td></td><td></td><td></td><td></td><td>27</td><td>72</td><td>26</td><td>69</td><td></td><td></td><td></td><td></td></tr>
<tr><td></td><td></td><td></td><td></td><td>30</td><td>65</td><td>31</td><td>68</td><td></td><td></td><td></td><td></td></tr>
<tr><td></td><td></td><td></td><td></td><td>55</td><td>44</td><td>54</td><td>41</td><td></td><td></td><td></td><td></td></tr>
<tr><td></td><td></td><td></td><td></td><td>50</td><td>45</td><td>51</td><td>48</td><td></td><td></td><td></td><td></td></tr>
<tr><td></td><td></td><td></td><td></td><td>43</td><td>56</td><td>42</td><td>53</td><td></td><td></td><td></td><td></td></tr>
<tr><td></td><td></td><td></td><td></td><td>46</td><td>49</td><td>47</td><td>52</td><td></td><td></td><td></td><td></td></tr>
</table>

Each face of this magic cube adds up to 194.

Further shapes (e.g. stars, polyhedrons, triangles) may be found in:

-    "Mathemagic" by Royal Vale Heath

-    "Mathematical Carnival" by Martin Gardner.

4.   Reversible Magic Square

This is an unusual magic square, in that it sums to a "magic total" of 264, even if you turn the magic square upside down!

96	11	89	68
88	69	91	16
61	86	18	99
19	98	66	81

Further details may be found in:

-    "Self-Working Number Magic" by Karl Fulves

-    "Greater Magic" by John Northern Hilliard.

5.   "Anti-Magic" Square

The following is an interesting variation on the 3 x 3 magic square, also from Karl Fulves's book "Self-Working Number Magic", where the aim is to produce a square that has as many **different** totals as possible:

5	1	3
4	2	6
8	7	9

It can be seen that each of the three rows, each of the three columns, and each of the two corner diagonals produces a different total, namely, in ascending order, 9, 10, 12, 13, 16, 17, 18 and 24.

Further "anti-magic" squares, such as the following one which produces totals of 6, 12, 15, 16, 17, 18, 19 and 21, may be found in "The Magic Numbers Of Dr. Matrix" by Martin Gardner:

1	2	3
8	9	4
7	6	5

6.   **"Diabolic" Squares**

As stated earlier, "diabolic", or "panmagic" squares are those where the "broken" diagonals also add up to the "magic total".

However, in addition to this property, a "diabolic" magic square remains a magic square even if a row is moved from the top to the bottom or from the bottom to the top, or if a column is moved from the left to the right or from the right to the left. This is the "diabolic" magic square found in Khajuraho, India:

7	12	1	14
2	13	8	11
16	3	10	5
9	6	15	4

Notice that the "broken" diagonal using the numbers 12, 2, 5 and 15 adds up to the "magic total" of 34.

If the bottom row is moved to the top:

9	6	15	4
7	12	1	14
2	13	8	11
16	3	10	5

the same "broken" diagonal, this time using the numbers 6, 7, 11 and 10, **still** adds up to the "magic total" of 34.

Similarly, if the left column is now moved to the right:

6	15	4	9
12	1	14	7
13	8	11	2
3	10	5	16

the same "broken" diagonal, this time using the numbers 15, 12, 2 and 5, **still** adds up to the "magic total" of 34.

A further property of "diabolic" magic squares is that, if you create a "mosaic" by fitting together a number of duplicate "diabolic" squares, then any four adjacent cells on this "mosaic" will sum to the "magic total":

7	12	1	14	7	12	1	14	7	12	1	14
2	13	8	11	2	13	8	11	2	13	8	11
16	3	10	5	16	3	10	5	16	3	10	5
9	6	15	4	9	6	15	4	9	6	15	4
7	12	1	14	7	12	1	14	7	12	1	14
2	13	8	11	2	13	8	11	2	13	8	11
16	3	10	5	16	3	10	5	16	3	10	5
9	6	15	4	9	6	15	4	9	6	15	4
7	12	1	14	7	12	1	14	7	12	1	14
2	13	8	11	2	13	8	11	2	13	8	11
16	3	10	5	16	3	10	5	16	3	10	5
9	6	15	4	9	6	15	4	9	6	15	4

as can be seen in the above example.

7. **Latin Squares & Graeco-Latin Squares**

Latin Squares are formed with "n" sets of the numbers from one to "n", where "n" is the number of cells on each side of the square. They should be arranged so that no horizontal row or vertical column contains the same two numbers.

If the two corner diagonals also contain no repeated numbers, then it is called a **diagonal** Latin Square. By definition, therefore, a diagonal Latin Square must have a "magic total" common to all rows, columns and corner diagonals, because the same numbers are used on each row, with no numbers being repeated in any row, column or corner diagonal.

Latin Squares may also be composed using letters - note that the square on the title page is such a square - or any other symbols (e.g. ESP symbols, road signs).

Graeco-Latin Squares (also known as Euler Squares) are similar, except that each cell contains a two-digit number, such that the tens digits and the units digits independently form Latin Squares, as shown in the following example:

11	22	33	44
34	43	12	21
42	31	24	13
23	14	41	32

See "Mathematical Recreations" by Maurice Kraitchik for further details.

8.   **Dominoes**

Magic squares may also be formed using dominoes. The following one, for example, uses nine dominoes to create a 3 x 3 magic square whose "magic total" is 15 (but note that both sets of "dots" on each domino must be added together to form a single number):

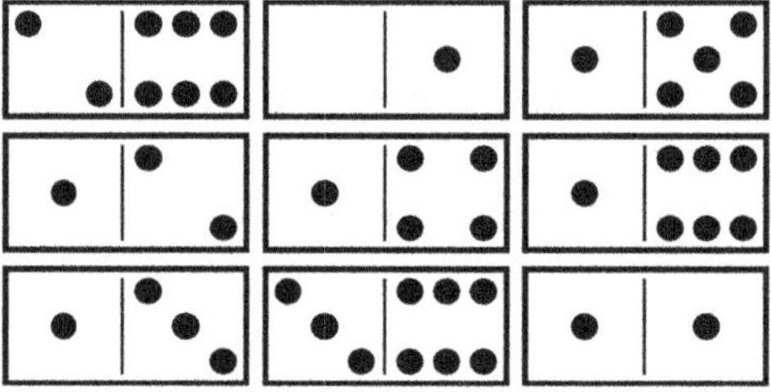

The following 4 x 4 magic square, whose "magic total" is 18:

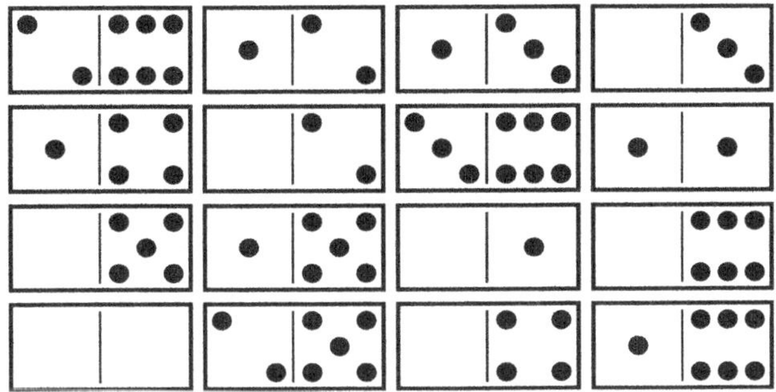

and 5 x 5 magic square, whose "magic total" is 27, should be treated as per the above 3 x 3 magic square, but note that, although no dominoes are duplicated, their totals are (e.g. there is a three/five domino and a six/two domino, both of which total eight):

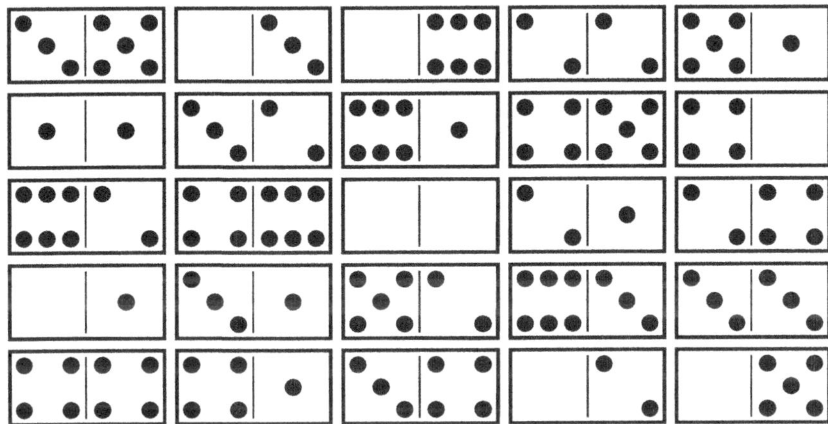

The 6 x 6 magic square is the largest solid square that may be formed using a double six set. The following 18 dominoes may be used to form such a magic square, whose "magic total" is 13, if each half of each domino is treated as a single digit in its own right:

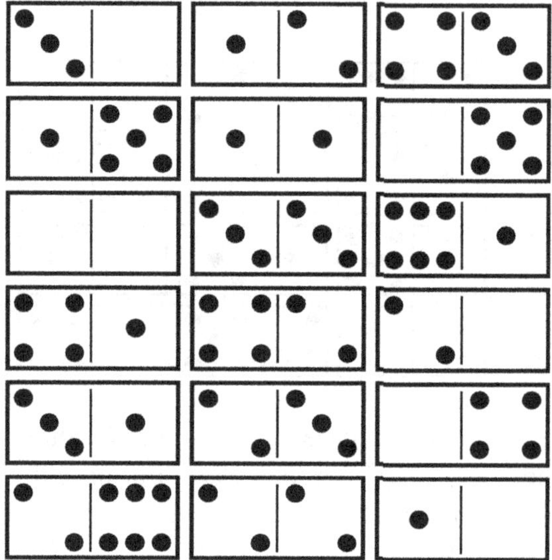

If the right most column is ignored, the following 28 dominoes (see next page) may be used to form a 7 x 7 magic square, whose "magic total" is 24, if each half of each domino is treated as a single digit in its own right:

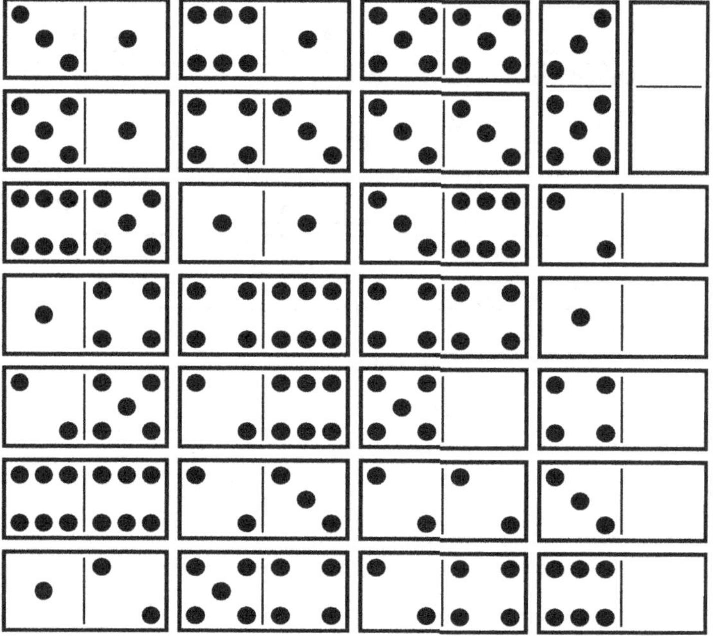

Further details may be found in:

- "Mathematical Recreations" by Maurice Kraitchik

- "Solo Games" by Gyles Brandreth.

9. **Subtraction / Multiplication / Division Magic Squares**

Although magic squares are usually treated "additively" (i.e. each number is added to form a "magic total"), by swapping the two pairs of corner numbers in a "basic" 3 x 3 magic square, a magic "subtraction" square is formed:

2	1	4
3	5	7
6	9	8

The "magic total", which is five, is obtained by summing the numbers at each end of a row, column or diagonal, and then subtracting the number from the middle cell (e.g. 6 added to 8 is 14, less 9 gives 5).

The following magic square is a magic "multiplication" square:

12	1	18
9	6	4
2	36	3

The "magic total", which is 216, is obtained by multiplying the three numbers in any row, column or diagonal. This is the simplest magic "multiplication" square, and produces the lowest possible "magic total", assuming that each cell contains a different positive integer.

By applying the same transformation to this magic "multiplication" square that was applied to the "basic" magic square, above (i.e. swapping the two pairs of corner numbers), a magic "division" square is formed:

3	1	2
9	6	4
18	36	12

The "magic total", which is six, is obtained by multiplying the numbers at each end of a row, column or diagonal, and then dividing by the number from the middle cell (e.g. 3 multiplied by 18 is 54, divided by 9 gives 6).

Further details may be found in "The Magic Numbers Of Dr. Matrix" by Martin Gardner.

10.　　**"Emirp" Squares**

An "emirp" is, apart from being the word "prime" backwards, a name coined by Joseph Farrell for a prime number that is not palindromic but does produce a different prime number when its digits are reversed.

Another mathematician, Leslie Card, has used these "emirps" to generate "emirp" squares, where each row, column and corner diagonal is a different "emirp".

The following 4 x 4 "emirp" square is unique, in that it is the only one possible:

9	1	3	3
1	5	8	3
7	5	2	9
3	9	1	1

"Emirp" squares of order five may also be created, although there is more than one such square that may be produced.

For further details, see "The Magic Numbers Of Dr. Matrix" by Martin Gardner.

11. **A Mathematical Problem**

There is a reasonably well-known mathematical problem, which is usually stated as follows:

> What is the only nine-digit number, that contains all of the digits from one to nine once only, whose first two digits (from left to right) are evenly divisible by two, and whose first three digits are evenly divisible by three, and whose first four digits are evenly divisible by four, and so on, up to and including all nine digits which are evenly divisible by nine?

Surprisingly, the solution may easily by found from a "basic" 3 x 3 magic square, by following the three arrows, starting with the one at the top left and finishing with the one at the bottom right:

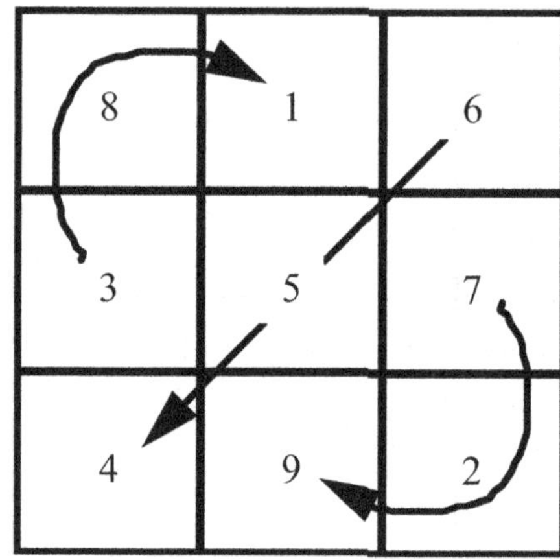

This generates the number 381,654,729, which is the unique solution to this puzzle. Further details and analysis may be found in "The Magic Numbers Of Dr. Matrix" by Martin Gardner.

9.   Presentation / Routines

The following presentations and routines all involve the use of magic squares, or utilise principles based on magic squares:

1.   Creating A Magic Square Which Sums To A Chosen Number

The most obvious presentation of a magic square is to create one which adds up to a number chosen by a spectator. The essence of performing this feat is to be able to complete the required magic square as quickly as possible, and without reference to any aide-mémoires. See Chapter 4, General Formulae, above for details of how to create a magic square adding up to a specified "magic total".

Whilst this can clearly be performed with any size of magic square, the best size to use is either a 4 x 4 magic square or a 5 x 5 magic square. Whilst a 3 x 3 magic square is both quick and easy to create, it is too small, and there are therefore too few combinations which add up to the selected number. However, 4 x 4 magic squares and 5 x 5 magic squares are:

-        not too large

-        are still reasonably simple and quick to create

-        have a sufficiently large number of combinations which add up to the selected number to be impressive.

Larger magic squares can be used, e.g. a 7 x 7, but are unnecessary.

Ideas and routines for this demonstration can be found in many books, some of which are:

-        "Feature Magic For Mentalists" by Will Dexter

-        "Magical Fun With Magic Squares" by Dr. Frank Blaisdell

-        "Mathemagic" by Royal Vale Heath

-        "Mathematical Magic" by William Simon

-        "Paul Daniels' Adult Magic" by Barry Murray

-        "Self-Working Number Magic" by Karl Fulves

-        "The Magic Book" by Harry Lorayne

-       "13 Steps To Mentalism" by Tony Corinda.

2.   **Creating A Magic Square That Starts At A Chosen Number**

A similar demonstration to the above is where a magic square, of any size, is created that starts with a number chosen by a spectator. As soon as the number is given, the performer writes down and hides a prediction. After the magic square has been completed, using the number chosen by the spectator as the lowest number, the performer shows that the magic square adds up to the "magic total" predicted earlier.

The effect is accomplished using the formula given in Chapter 4, General Formulae, above.

3.   **"Window Screens"**

In his book "Magical Fun With Magic Squares", Dr. Frank Blaisdell introduces the idea of what he calls "window screens". These are large cards with cut-out windows which allow the performer to show each combination of four squares which sum to the required total.

However, as the author freely admits, he is only aware of 30 such combinations. As has been stated earlier in Chapter 6, An Analysis Of 4 x 4 Magic Squares, of this booklet, there are in fact up to **86** combinations that add up to the chosen number, a fact that probably makes it impractical to show **all** combinations using these "window screens". However, as a means of showing some of the more obvious ones, then it is an excellent idea, but remember that the combinations that do add up to the required "magic total" vary with the remainder, so you must pick combinations that are common to all magic squares, regardless of the remainder.

4.   **Memory Pack/Magic Square Demonstration**

This demonstration, which is a routine of my own, combines a demonstration of memorising a pack of cards and the creation of a magic square.

Effect

The performer shows a shuffled, full pack of playing cards, and gives a third of them to a member of the audience to shuffle and distribute to people sat nearby. A second third is also given out for shuffling and distribution.

A third spectator is given an envelope and asked to choose a card from the remaining third. This card is placed, unseen by anybody, in the envelope which this third spectator holds for a few minutes. This last third is also shuffled and distributed to people sat nearby.

The performer then points out that each card has a random number written on its face. Members of the audience are asked to call out a number, and the performer replies with the name of the card.

After a few such tests, the performer introduces a blackboard, which has been marked into 16 squares, each lettered from "A" through to "P".

Members of the audience are asked to call out one of the letters on the blackboard, and the performer then asks the person who is holding a card nominated by the performer to shout out the number written on that card. The stated number is written in the requested cell. This process continues until each of the 16 cells has been completed.

The performer then turns to the spectator who has been holding the envelope and asks him/her to open it for the first time and name the card inside. Having done this, he/she is also asked to call out the number written on that card.

The performer then points to the blackboard and demonstrates that a magic square has been created which adds up to the hitherto unknown number in up to **86** different ways.

Properties & Set-Up

A full pack of playing cards is required which needs to be arranged into a known order. Systems such as the Nikola card system, the Aronson Stack, or any other "memory pack" will do - suitable systems can be found in:

- "Bound To Please" by Simon Aronson

- "The Memory Book" by Harry Lorayne & Jerry Lucas

- "How To Develop A Perfect Memory" by Dominic O'Brien - whose techniques **must** work, since he can remember the sequence of a staggering **35** packs of shuffled cards!

- "Memory Of The Mind" by Eddie Joseph.

Note that other well-known arrangements, such as the Si Stebbins or Eight Kings set-ups, will only suffice **provided** that you know the position of each card instantly.

A matching Joker is also required.

Each card in this pack is then numbered on its face according to its position in your personal memory pack, with the exception of the Joker.

The cards numbered from 1 to 33, and also the Joker, are shuffled.

The card numbered 34 should be cut short, to facilitate the cutting of the pack into thirds. The cards numbered from 34 to 52 are placed, in sequence, with 34 at the top and 52 at the face, underneath the shuffled cards.

A small envelope, a marked-off blackboard and some chalk, as detailed above under Effect, are also required.

Finally, the performer must be able to create a 4 x 4 magic square, adding up to a chosen total, using the formula detailed above in Chapter 4, General Formulae.

Routine

The pack is removed from its case and fanned, concealing as much of the faces as possible to disguise the fact that one third of the deck is in numerical sequence.

One third of the pack is cut off and given to a spectator for shuffling and distribution.

The second third is cut off, using the short card (i.e. the card numbered 34) to cut to the correct place. This section is also handed out for shuffling and distribution.

The envelope is given to a third spectator, and the remaining cards are slowly spread, face down, from hand to hand. When the spectator says "Stop", the selected card is slid, unseen, into the envelope, and the cards are cut at that point.

By glimpsing the card that is now on the face of this remaining section of the pack, it is easy to determine the card that was selected; the total required for the magic square will be one greater than the card now on the face of this packet. This last section is now shuffled, to hide the evidence, and given out for distribution.

The performer now points out that the cards are apparently randomly numbered, with no two cards sharing the same number, and asks for numbers to be called out. Using the "memory pack" system, the performer should have no difficulty in achieving this. There may, however, be two exceptions:

1. A spectator calls out the name of a card instead of the number written on that card. In this case, the performer should respond, say, with "No, Sir/Madam, you should have called out number 'x', and then **I** would have told you that you are looking at the 'value' of 'suit'". This obviously means that you must be able to determine the card at each position, **and** the position for each card.

2. A spectator calls out that their card has no number on it. This will be the Joker, therefore the performer should respond, say, with "There's always at least one Joker, and that is the card you are looking at, is it not?"

After a few numbers have been called out, the performer moves on to the second half of the demonstration.

Spectators call out squares from the blackboard. When each square is named, the performer must:

1. Determine which number should be written in that cell to form a magic square which will add up to the number on the unseen card in the envelope.

2. Substitute and call out the correct playing card, according to the memory pack being used, for the number calculated in Step 1.

When the entire square has been completed, the performer should have the envelope opened, and its playing card and number called out. The performer then shows that a magic square has been created, adding up to the hitherto unknown number in up to **86** different ways. (See Chapter 6, An Analysis Of 4 x 4 Magic Squares, above, and Appendix A, Total Combinations Of Four Squares Which Add Up To A Selected Total - Complex Formula, for further details.)

Variations

The card which bears the number representing the magic square's "magic total" could, of course, be forced, in which case:

- the card that is numbered 34 need not be cut short

- the entire pack could be shuffled (i.e. no stack is required), depending on the type of force preferred.

Acknowledgements

The idea of performing a memory test using a pack of cards in this way is Joe Riding's. My contributions are:

1. Including the Joker, with no number written on it, which leads to an opening for a light-hearted "gag" about "there always being one Joker".

2. Using the memory pack as a means of creating a magic square.

Joe Riding has recently published an alternative routine for presenting magic squares, called "Computer Error", which is a combination of Creating A Magic Square That Sums To A Chosen Number, described above, and this effect. Joe points out that, in the traditional presentation of magic squares, it becomes obvious to the audience during the effect that a magic square is being created, since the numbers are written in the cells as the effect progresses. Joe's routining prevents this from happening.

5. **The Amazing Magic Square & Master Memory Demonstration**

This effect, created by the late Orville Meyer, is one of my favourite routines, as it combines the creation of a magic square with a memory demonstration. Experience has shown that the appeal of this effect is twofold: the memory demonstration appeals to some, whilst the creation of the magic square appeals to others.

The basic effect is that a number (from 34 to 100) is called out by a member of the audience. A blackboard which has been marked off into 16 squares, labelled from "A" through to "P" is then completed, in a random sequence, partly by the performer and partly by the audience:

-        a square is called out by a spectator (e.g. "A")

-        the performer then calls outs a number which is written in the nominated cell (e.g. 17)

-        a member of the audience then calls out any object, which is also written in that cell (e.g. table).

This is continued until all 16 squares have been completed.

Members of the audience may then call out any of the three elements in any cell, i.e. the letter, the number or the object. The performer instantly responds with the other two elements. For example, if the number 16 is called out, the performer might reply with letter "K" and "Gentleman's Wallet".

After a few such tests, the performer quickly repeats all of the objects, in reverse order (i.e. from square "P" back to square "A").

Finally, the performer shows the audience that, not only has he memorised all 16 objects, but **also** he has created a magic square which adds up to the number, chosen at the start of the demonstration, in up to **86** different ways.

And, of course, all of the above can be performed entirely blindfolded!

This following example, taken from one of my own performances in 1995 when the Northamptonshire Magicians' Club put on a show for the Leamington & Warwick Magic Society, is what a square might look like at the end of the demonstration:

A 17 Table	B 14 Record	C 11 Juggling Balls	D 27 Bow Tie
E 12 CD	F 26 Hyperbolic Paraboloid Roof	G 18 Star	H 13 Cod Fish
I 25 Key	J 9 Guide Dog	K 16 Gentleman's Wallet	L 19 Picture
M 15 Perfume (Beauty)	N 20 Baby's Bottle	O 24 Pair Of Glasses	P 10 Coat

The key to this effect is a series of mnemonics for the "basic" magic square. Each mnemonic encodes both the square's letter (e.g. "A") and the number contained in that cell in the "basic" magic square (e.g. 9). The rest of the method relies on:

- a formula for amending this "basic" magic square so that the spectator's chosen number is produced (see Chapter 4, General Formulae, above)

- standard associative mnemonic techniques as used by many mentalists. Methods for this can be found in many books, including:

  - "A Question Of Memory" by David Berglas & Guy Lyon Playfair

  - "How To Develop A Perfect Memory" by Dominic O'Brien

-        "Instant Memory - The Way To Success" by Robert Harbin

-        "The Memory Book" by Harry Lorayne & Jerry Lucas

-        "The Unforgettable Memory Book" by Nick Mirsky

-        "You Can Remember" by Dr. Bruno Furst.

The full method used to achieve this effect is fully described in Orville Meyer's 18 page booklet, "The Amazing Magic Square And Master Memory Demonstration".

Below is a 5 x 5 magic square which I devised that could be used for an expanded version of the above routine:

A 23	B 6	C 19	D 2	E 15
F 4	G 12	H 25	I 8	J 16
K 10	L 18	M 1	N 14	O 22
P 11	Q 24	R 7	S 20	T 3
U 17	V 5	W 13	X 21	Y 9

Note that this magic square is the "transposed" version of the "basic" 5 x 5 magic square, shown above in Chapter 7, An Analysis Of 5 x 5 Magic Squares.

The mnemonic key words that I have developed for use in this expanded version of Orville Meyer's effect are as follows:

Square	Value	Mnemonic	Notes
A	23	Anaemia	Imagine the named object having all colour drained out of it.
B	6	Bush	Imagine a large bush, with lots of the named object growing on it.
C	19	Cut Up	Imagine the named object being cut to shreds with a pair of scissors or shears.
D	2	Don	Imagine the named object wearing a mortar board and gown like a University Don.
E	15	Ethyl	Imagine the named object getting drunk on a bottle of ethyl alcohol.
F	4	Fire	Imagine the named object bursting into flames.
G	12	Godown	Imagine a warehouse being stacked high with the named object.
H	25	Henley	Imagine the named object on a boat at the Henley Regatta.
I	8	Ivy	Imagine the named object growing all over the side of a house, in place of ivy.
J	16	Judge	Imagine a judge banging the named object, instead of his gavel, on the bench.
K	10	Kites	Imagine the named object(s) flying high in the sky like kites.

Square	Value	Mnemonic	Notes
L	18	Lot Of	Imagine a very large heap of the named object(s).
M	1	Mat	Imagine the named object sitting on a doormat, instead of the cat (as in "the cat sat on the mat").
N	14	Nutter	Imagine the named object wearing a straitjacket, like a nutter in a lunatic asylum.
O	22	Onion	Imagine peeling the outer layers off the named object, as you would with an onion, and it making you cry.
P	11	Potted	Imagine the named object being potted in a plant pot, instead of the plant.
Q	24	Queen ER	Imagine Queen Elizabeth wearing the named object on her head, instead of her crown.
R	7	Rake	Imagine sweeping up the leaves in your garden using the named object instead of a rake.
S	20	Sneeze	Imagine sneezing and the named object shooting at great speed out of the end of your nose.
T	3	Time	Imagine the named object telling the time instead of a clock.
U	17	You (U) Dog	Imagine the named object doing a James Cagney impersonation, saying "You (dirty) Dog!"
V	5	Veil	Imagine the named object wearing a veil, or yashmak, like a Muslim lady.

Square	Value	Mnemonic	Notes
W	13	With Me	Imagine that you are walking along the sea front, hand in hand with the named object instead of with your loved one.
X	21	Xanadu	Imagine a gold-plated, diamond-studded version of the named object in Kubla Khan's luxurious pleasure palace.
Y	9	Yob	Imagine a yob, or lager lout, spray-painting a huge picture of the named object on a wall.

Note that:

1.   The mnemonic system for remembering numbers that I use is the most common "figure alphabet", which was developed by Stanislaus Mink von Wennssheim in the mid-seventeenth century. Further work on this technique was published in a treatise by Gregor von Feinaigle of Baden in 1812. In this system, the following letter or letters are used to represent each digit:

Digit	Letter(s)
0	C, S or Z (all "soft" sounds)
1	D or T
2	N
3	M
4	R
5	L
6	CH, G, J or SH (all "soft" sounds)
7	C, G or K (all "hard" sounds)
8	F or V

Digit	Letter(s)
9	B or P

Further details may be found in the following books:

- "A Question Of Memory" by David Berglas & Guy Lyon Playfair

- "Dynamic Mentalism" by Stanton Carlisle

- "Instant Memory - The Way To Success" by Robert Harbin

- "The Memory Book" by Harry Lorayne & Jerry Lucas

- "The Unforgettable Memory Book" by Nick Mirsky.

Similar systems are detailed in:

- Step 3 of Tony Corinda's "13 Steps To Mentalism"

- Eddie Joseph's "Memory Of The Mind"

- "You Can Remember" by Dr. Bruno Furst.

2. As with all suggested mnemonics, remember that these are my own words; you should always try to find your own, since they will be more personal to you, and, therefore, more memorable.

3. Even if you don't learn or know a full mnemonic system, if you learn one of the above two magic square and memory demonstrations, you will still be able to present a 16 or 25 object memory test, using the keywords from the relevant magic square.

6. **The Marrakech Game**

This is an interesting game which **apparently** has no connection to magic squares.

The numbers from 1 to 9 are written down, and then two players take turns in calling out one of the numbers that has not yet been used. The numbers called out by the "sucker" are circled, whilst those called out by the performer have squares drawn around them. The winner is the first person to call out three numbers that add up to 15.

Needless to say, the performer never loses! How? It's simple - unknown to the "sucker", the performer is actually constructing a 3 x 3 magic square!

Full details can be found in Karl Fulves's book, "Self-Working Number Magic".

7.   **The Tit-Tat-Toe Game**

This game is similar to the Marrakech Game, above, except that a game of tit-tat-toe (or noughts and crosses) is played using playing cards. The Ace to nine of one suit are used, apparently in the order in which they are found in a shuffled deck. The performer lays his cards face down and the spectator plays his cards face up.

The game ends in a draw, but when the performer turns his cards face up, it is seen that a "basic" 3 x 3 magic square has been created, whose "magic total" is 15.

The basic set-up for this effect is that the Ace to nine of one suit should be ordered as follows, from the top to the face: Ace, eight, two, seven, three, four, five, six, nine. This stack is then placed at the bottom (i.e. face) of the pack. The pack is riffle shuffled twice, and then the nine cards are removed from the pack in the sequence in which they are found. Since the riffle shuffle will not alter this sequence, they will be in the order required.

With the magician starting, by placing his first face down card in the centre square, there are two different strategies to follow, depending on whether the spectator plays his first face up card in a corner square or "middle" square.

I have devised an alternative set-up, whereby nine numbered cards may be used in isolation. The cards must be placed in numerical order, with the Ace (or one) at the face and the nine at the top. The cards are then shown individually, from hand to hand, reversing the sequence at the same time. The cards are then turned face down and mixed as follows:

1.   Take the bottom card of the packet into the other hand.

2.   Take the top card of the packet and place it on top of the first card taken.

3.   Take the new bottom card of the packet and place it on top of the first two cards taken.

4.   Take the new top card of the packet and place it on top of the first three cards taken.

5.   Take the new bottom card of the packet and place it on top of the new first four cards taken.

6.   Take the new top card of the packet and place it on top of the first five cards taken.

7.   Take the new top card of the packet and place it on top of the first six cards taken.

8.   Take the new top card of the packet and place it on top of the first seven cards taken.

9.   Take the last card of the packet and place it on top of the first eight cards taken.

Then the cards should be shuffled, by "milking" the bottom card (which should be the nine) and the top card (which should be the six), then running off all other cards individually. This should place the nine cards in the required sequence, as per the original effect.

Full details can be found in either:

-       "Mathematical Magic" by William Simon

-       "Mathematics Magic And Mystery" by Martin Gardner.

8.   **Numorator**

This is an effect by Al Smith, in his book "Cards On Call", which has been used by Paul Daniels on one of his television shows. It was originally published in *Abracadabra* under the title "Only Sixteen".

A spectator completes a 4 x 4 grid, using consecutive numbers, starting from a number of his choice in the top left cell and finishing with his chosen number plus 15 in the bottom right cell. (This number may be merely thought of, or it may be selected by rolling several "real" or "imaginary" dice.) Whilst the spectator is doing this, the performer completes a similar square, using the same numbers but in an apparently random order.

The spectator then selects four numbers from his own square, using an elimination technique, and adds up their total. It is found that the square completed by the performer is in fact a magic square that adds up to the freely chosen total!

This is an excellent and simple routine, full details of which may be found in the above-mentioned book.

The principles behind this type of "magic square" are discussed more fully in "Mathematical Puzzles And Diversions" by Martin Gardner, and the elimination technique was originated by Mel Stover as a calendar prediction.

Three similar routines are detailed in "Card Bored?" by Stephen Tucker:

-       "The Square Comes Full Circle"

        This is a "dressed-up" version of the first part of "Numorator", described above and which was the inspiration for this routine, where the spectator completes a 4 x 4 grid, and the final elimination phase.

-       "The Almost Magic Square"

        In this routine, a spectator completes the first two rows of a 4 x 4 grid using two sets of four freely chosen consecutive numbers. The magician then writes down an instruction on a piece of paper. The spectator then completes the third row of the grid using four more consecutive numbers, and finally completes the fourth row following the instructions previous written by the magician.

        The completed grid forms a "partial" magic square, to which the elimination technique, described above, may also be applied.

-       "Sweet Sixteen"

        This is a simple but impromptu version of "The Square Comes Full Circle", described above.

9.   **The Paradox**

This is another routine from Karl Fulves's book, "Self-Working Number Magic". It is a combination of a magic square and a five piece puzzle. Initially, there are five pieces of paper which have numbers written on them which are formed into a square; it is seen that it is in fact a 3 x 3 magic square which adds up to 45 in all eight directions.

The pieces are all turned over, and the spectator is requested to form another magic square. He will find that he only needs four of the five pieces to form this new magic square; the fifth, unused piece, when turned over, reveals the "magic total" of this new square (which is 15).

10.   **Astro Thought**

This is another effect from Karl Fulves's "Self-Working Number Magic", where nine cards are used, each having a number from one to nine on one side and one of three different colours on the other.

The cards are placed number side down on the table and mixed up. Three spectators each choose three of the cards, one of each colour.

The three spectators then decide on an order for the three colours, e.g. red, yellow then blue.

Each spectator in turn calls out the number on each of his three cards, in the order specified, to form a three-digit number, which is written down by the performer.

These three numbers are totalled and seen to match a prediction made earlier by the performer. This is because this will always force the number 1,665, since this is what a 3 x 3 "basic" magic square will always sum to if you add up each row as a three-digit number.

Full details of the effect can be found in the above-mentioned book, and a routine using this principle as a force of the number 1,665 may also be found in an effect called "To Hell And Back" in Roy Johnson's "Unique" lecture notes.

11.   **Déjà Vu**

This is a similar effect to Astro Thought, above. It may be found on "The Larry Becker" lecture video, produced by Davenports. This version, however, has **two** predictions.

Nine cards, numbered from one to nine, are shown, shuffled and then distributed, face down, to three members of the audience.

One of these three spectators, who is freely chosen by a fourth spectator, is given an envelope to hold.

The remaining two spectators then alternately call out numbers from their cards, thus forming three two-digit numbers, which the performer writes down.

The person holding the envelope adds up the total of his three cards, opens the envelope and finds that it contains a prediction of his total.

The three two-digit numbers formed at random by the other two spectators are then added up, and their total is also shown to match a prediction made earlier.

By removing one set of three cards, this will force both the number 15 (i.e. the set of three cards) and the number 165 (i.e. the sum of the two remaining sets of cards).

Full details may be found in the above-mentioned video.

12.   **"15" Force**

The magic square principle has been used by Max Maven to force the number 15 (see the October 1985 edition of "The Linking Ring" for details).

A pack of cards is shuffled and the top eight cards are dealt into four piles of two cards. The next card is given to a spectator, who places this ninth card on top of any of the pairs; the other six cards are returned to the pack. When the spectator turns over the remaining pile of three cards and adds up the total, it is found that the sum is 15.

This fact may be used in whatever way the magician desires; at its simplest, it could be a straightforward prediction. A different routine using this principle may be found in the July 1995 edition of "The Linking Ring".

The force is based on the fact that a 3 x 3 magic square consists of four pairs of diametrically opposite numbers that add up to ten, with the central square containing a value of five.

(It may be proved that the centre cell of a "basic" 3 x 3 magic square **must** contain the number five; see "Mathematical Recreations" by Maurice Kraitchik or "In Code" by Sarah & David Flannery for details.)

13.   **The Birthday Square I**

This is a routine which builds a 3 x 3 magic square based on a spectator's date of birth.

During the creation of this magic square, the spectator may choose when to use the "day" and when to use the "month", and whether these should be added to or subtracted from other numbers already entered.

For example, a magic square created using my own date of birth, 30/5/59, may look like this:

114	59	79
49	84	119
89	109	54

This particular magic square, in which the Day was added then the Month was subtracted, adds up to a "magic total" of 252. The finished magic square may be given to the spectator, preferably on the back of one of your business cards, as a souvenir, stating that the "magic total" is their lucky number, and that it is also lucky to carry the magic square with them at all times.

The original version of this routine, kindly given to me by Alan Shaxon, was published by Les Vincent in the *Gen* magazine. In this version, the day, month and year figures are only added to each other in various combinations.

The above version, whereby the day and month may be added or subtracted, is fully detailed in:

-        "Mathematical Magic" by William Simon

-        "Greater Magic" by John Northern Hilliard.

However, further analysis, over and above that shown in the above books, reveals that there are only eight different ways in which this type of magic square may be constructed:

1. Add the Day then add the Month.
2. Add the Day then subtract the Month.
3. Subtract the Day then add the Month.
4. Subtract the Day then subtract the Month.
5. Add the Month then add the Day.
6. Add the Month then subtract the Day.
7. Subtract the Month then add the Day.
8. Subtract the Month then subtract the Day.

Note, however, that the first four options produce the same "magic total" as the second set of four options (i.e. it does not matter in which order the Day and Month are used).

The "magic total" may easily be calculated as the result of (Year +/- Day +/- Month) multiplied by three. In the example above, where the Day was added then the Month was subtracted, this total is, as shown above, (59 + 30 - 5) multiplied by three, giving 252. Therefore, if the spectator gives his options of whether to use the Day or Month first, and which should be added or subtracted, before anything is written in the magic square, this formula may be used to write down a prediction of what the "magic total" will be.

Note that, depending on the values of the Year, Month and Day, the magic square may contain negative numbers if subtraction is selected for either or both of the Month and Day. As a guideline, the smallest number that will ever appear in the magic square is (Year - twice the Day - twice the Month). As an example, in the above magic square that uses my date of birth, subtracting the Day would result in negative values.

However, as John Northern Hilliard points out in "Greater Magic", if the entire Year is used (i.e. including the century) instead of just the last two digits, then this problem would never arise. Creating the magic square using this approach with my "full" date of birth, 30/5/**1959**, enables both the Day and the Month to be subtracted, if required:

1894	1959	1919
1949	1924	1899
1929	1889	1954

Using the formula given above, it may be seen that the "magic total" of this magic square is (1959 - 30 -5) multiplied by three, giving 5,772. This may, of course, be checked by adding up the different rows, columns and corner diagonals in the above magic square.

An obvious but minor disadvantage of this approach (i.e. using the full Year) is that larger numbers have to be manipulated. This should not be a problem with the addition and/or subtraction of the Day and Month values, but it does make it more difficult to determine, and optionally predict, the "magic total" before the magic square has been completed.

13.   **The Birthday Square II**

Here is another variation on a magic square built using somebody's birthday.

It was while I was discussing magic squares with David Berglas, who was the President of The Magic Circle at the time, that he told me about a routine he had performed whereby a person's birthday was included within four cells of a 4 x 4 magic square. (David's Magic Square routines are detailed in the excellent book by David Britland, "The Mind & Magic of David Berglas".)

What follows is the method that I eventually discovered of creating such a Birthday Magic Square, where the top row of four cells forms a specific date (e.g. somebody's birthday, anniversary, or perhaps Christmas Day), and it was originally published in M.U.M., which is the magazine of the Society of American Magicians.

Effect

A 4 x 4 magic square is created, with the four cells in the top row forming a specific date.

Method

There are two basic parts to the method: the first is knowing a "standard" 4 x 4 magic square, and the second is knowing how to adjust the cells in this standard magic square so as to create the Birthday Magic Square.

The standard 4 x 4 magic square has already been covered in earlier chapters, and, ideally, this standard magic square should be memorised, although it can be written on some form of "crib sheet", depending on how and where you wish to use this effect.

For ease during the remainder of these instructions, the 16 cells will be referred to as A through to P, where A is the top left cell and P is the bottom right one.

The first step is to replace the four cells in the top row (i.e. 9, 6, 3 and 16) with the four components of the specified date. For this example, we shall use my own birth date of 30 May 1959, so the new top row would contain the numbers 30, 5, 19 and 59, so the partially built Magic Square would look like this:

30	5	19	59

The other twelve cells then need to be adjusted relative to the four cells in the top row.

As, in the standard magic square, Cell A contains the next to highest number in the first row, then the next to highest cell in each of the remaining three rows needs to be adjusted in by the same value. As Cell A usually contains 9, and it now contains, in this instance, 30, then the adjustment value is +21. Applying this adjustment to the other three rows means that: Cell G, which usually contains 10, now contains 31; Cell L, which usually contains 11, now contains 32; and Cell N, which usually contains 12, now contains 33.

Our partially built magic square now looks like this:

30	5	19	59
		31	
			32
	33		

Next, we move on to Cell B, which, in the standard magic square, contains the nest to lowest number in the first row. This means that the next to lowest number in each of the three remaining rows also needs to be adjusted by the same value. As Cell B usually contains 6, and it now contains, in this instance, 5, then the adjustment value is -1. Applying this adjustment to the other three rows means that: Cell H, which usually contains 5, now contains 4; Cell K, which usually contains 8, now contains 7; and Cell M, which usually contains 7, now contains 6.

Our partially built magic square now looks like this:

30	5	19	59
		31	4
		7	32
6	33		

Next, we move on to Cell C, which, in the standard magic square, contains the lowest number in the first row. This means that the lowest number in each of the three remaining rows also needs to be adjusted by the same value. As Cell C usually contains 3, and it now contains, in this instance, 19, then the adjustment value is +16. Applying this adjustment to the other three rows means that: Cell E, which usually contains 4, now contains 20; Cell J, which usually contains 1, now contains 17; and Cell P, which usually contains 2, now contains 18.

Our partially built magic square now looks like this:

30	5	19	59
20		31	4
	17	7	32
6	33		18

Finally, we move on to Cell D, which, in the standard magic square, contains the highest number in the first row. This means that the highest number in each of the three remaining rows also needs to be adjusted by the same value. As Cell D usually contains 16, and it now contains, in this instance, 59, then the adjustment value is +43. Applying this adjustment to the other three rows means that: Cell F, which usually contains 15, now contains 58; Cell I, which usually contains 14, now contains 57; and Cell O, which usually contains 13, now contains 56.

Our finished magic square now looks like this:

30	5	19	59
20	58	31	4
57	17	7	32
6	33	56	18

The "magic total" is, of course, the sum of the day, month, century and year, which is, in my case, 113.

Further Information

The effect above is a "bare bones" method. The routines in "The Mind & Magic of David Berglas" (see pages 457 - 469), by David Britland, are much more complete and are more suitable for "formal" performance than for impromptu use, which is what this particular magic square was intended for.

Acknowledgements

Although the method described above was my own creation, it undoubtedly is similar to the basic method used by David Berglas. However, the two were created independently, but I would certainly credit David with planting the seed in my mind to devise such a magic square.

14.   **The Sliding Square**

Many people may be familiar with the standard, sliding block puzzle, where 15 numbered blocks are arranged in a square tray in a random order, and the object is to slide the blocks into sequence, so that the number one is at the top left and the number 15 is at the next to bottom right (the bottom right being the empty space), as follows:

1	2	3	4
5	6	7	8
9	10	11	12
13	14	15	

This is not a particularly interesting puzzle in its own right, although it can be shown that exactly half of the possible arrangements are solvable and half are not (see "The Moscow Puzzles" by Boris Kordemsky for details).

In "The Moscow Puzzles" by Boris Kordemsky, the puzzle is turned around slightly: the object is **not** to slide the blocks into numerical sequence, but rather into a combination that forms a 4 x 4 magic square. This assumes that the empty space counts as zero, which means that any resulting magic square would have a "magic total" of 30 (using the formula given in Chapter 4, General Formulae, above).

Because of the two different types of starting position, there are two discrete types of magic square that may be formed. Further information may be found in the above-mentioned book.

15.   **The Knight's Tour**

The Knight's Tour is one of the oldest of knight puzzles on a chess board, and involves moving a knight, using standard rules of chess, so that it occupies each and every square on the board in turn, without moving to any square more than once.

The tour is deemed to be "closed" if the knight returns to its starting square, or "open" if it ends up on a different square to the one on which it started.

Although the Knight's Tour has no apparent link to magic squares, work was started on "magic" Knight's Tours by Euler in the eighteenth century, and a "semi-magic" square was published by William Beverley, in "The London, Edinburgh, and Dublin Philosophical Magazine and Journal of Science", for August 1848.

If each move is numbered, starting at one and ending at 64, then the pattern formed after an "open" tour creates a "semi-magic" square that adds up to a "magic total" of 260:

1	30	47	52	5	28	43	54
48	51	2	29	44	53	6	27
31	46	49	4	25	8	55	42
50	3	32	45	56	41	26	7
33	62	15	20	9	24	39	58
16	19	34	61	40	57	10	23
63	14	17	36	21	12	59	38
18	35	64	13	60	37	22	11

Note that the two corner diagonals do not add up to 260. It is this fact that prevents it being a full magic square.

Full details may be found in the "Mathematical Magic Show" by Martin Gardner.

16.   **"The Magic Square"**

This is a routine originated by the late Peter Kane, which may be found in his 1989 lecture notes, "Kane At Table".

The Ace through Nine of one suit are removed from a blue-backed pack of cards. These are arranged, face up on the table, to form a 3 x 3 magic square that adds up to 15 in the usual eight combinations. However, on turning the nine cards over (i.e. face down), it is seen that the back of each card is no longer blue, but has the number "5" printed on its back, each one being in a different colour. This is, therefore, another way of creating a "magic total" of 15.

Full details may be found in the above lecture notes. It should be noted, however, that the routine and patter may be altered to suit each performer by using different backs for the nine cards; for example, standard rainbow backs, a message that is spelled out on each card (e.g. "I-T'-S--M-A-G-I-C").

17.   **The Origami Magic Square**

This is dedicated to Neville "Pip" Ayris, a member of the Northamptonshire Magicians' Club for over 40 years who sadly and suddenly passed away in April 1996. Pip's other love apart from magic was origami, and it was my desire to find some means of combining my love of magic squares with Pip's love of origami. In the "Crossed Box Pleat", shown in "Origami" by Paulo Mulatinho but invented by Thoki Yenn, I believe that I have found that combination.

9	6			3	16
4					5
		15	10		
		1	8		
14					11
7	12			13	2

If the numbers shown above are written on a square piece of paper, and the "Crossed Box Pleat" fold is made, then a "basic" 4 x 4 magic square is formed.

Whilst it is difficult to depict on paper what this looks like, it could be described as a small 2 x 2 square superimposed on top of a 4 x 4 border, and this diagram indicates the idea:

9	6	3	16
4	15	10	5
14	1	8	11
7	12	13	2

Full instructions for creating the Crossed Box Pleat may be found in Appendix F.

18.   **"The Ultimate Magic Square"**

This is an effect that was invented by Christoph Wasshuber and is marketed by Mephisto Inc. in Belgium. The following is an abridged and combined version of the effect as described in the catalogue and the instructions:

The performer holds a packet of 16 number cards that represents a prediction. The spectator is told to imagine three "invisible" dice which he is to throw, calling out the sum of the numbers "showing" on all three dice. With his 16 cards, the performer produces a perfect magic square that has a "magic total" that sums to the completely freely chosen number. In addition, the 4 x 4 magic square formed by the 16 cards forms a message which reads "That's Magic!" Because this message is written over the entire magic square, it is obvious that the cards can be laid out in just one order. The 16 cards are never exchanged, and this miracle can be performed with the same 16 cards regardless of which numbers are "showing" on the three dice!

19.   **"The Number Of The Beast"**

This is a routine of my own.

Effect

The performer talks about the link between magic squares, black magic and the Devil, a link which dates back to Biblical days. After a hidden prediction is placed on the table, a spectator then chooses, by selecting a numbered card or piece of paper, a number from three to nine, and the selected number is written down.

The performer draws a "basic" magic square of an order determined by the chosen number. The spectator chooses any row or column from the magic square and finds the "magic total", which is written underneath the chosen number.

The spectator multiplies the chosen number by the "magic total", at which point the performer reveals a prediction made earlier, which is a quotation from the Bible, and which is proven to be correct:

> **Revelations, Chapter 13, Verse 18**
>
> "Here is wisdom. Let him that hath understanding count the number of the beast; for it is the number of a man, and his number is six hundred, threescore and six."

Properties & Set-Up

Seven cards are required, bearing the numbers three to nine inclusive. Note that playing cards, Tarot cards or Piatnik 1-2-3 cards may be used.

The back of the "six" card should be marked so that you can easily recognise it.

The cards should then be placed in an envelope.

If you wish to perform the effect impromptu, then the numbers one to nine may be written on a piece of paper, ensuring that the number six is written in the centre. By tearing the paper into nine pieces, the "six" piece will easily be recognisable, since it will be the only piece with four torn edges. Note that the pieces with the numbers 1 and 2 should be discarded, since magic squares may not be formed for orders 1 and 2.

The above prediction should be written, either on the back of the envelope in which the number cards are placed, or, for impromptu use, on a piece of paper inside an envelope.

Also required are a pen and a piece of paper.

Routine

After some appropriate patter, the prediction is laid on the table and the numbered cards or pieces of paper are spread, number side down, on the table.

The "six" is forced, using the P-A-T-E-O (Pick Any Two, Eliminate One) force. (Remember that, because there are an odd number of cards, the performer must start this process. I find that the mnemonic "the magician is odd" helps - after all, there is nobody "odder" than a magician!)

The "chosen" number, 6, is written on a piece of paper.

A "basic" 6 x 6 magic square (i.e. the one shown at the start of Chapter 7, Variations, above) is then drawn by the performer.

The spectator chooses any row or column and adds up the six numbers in that row / column. The result should be 111, since this is the "magic total" of a "basic" 6 x 6 magic square.

The number 111 is written underneath the number 6, and the two numbers are multiplied together to give a product of 666.

The performer then reveals the prediction.

Other Ideas

An alternative ending is to produce a "double" prediction, using the "Origami Magic Square" detailed above. The following diagram represents the unfolded piece of paper, showing seemingly unconnected numbers in the corners and middle of the paper, with the biblical quotation in the centre:

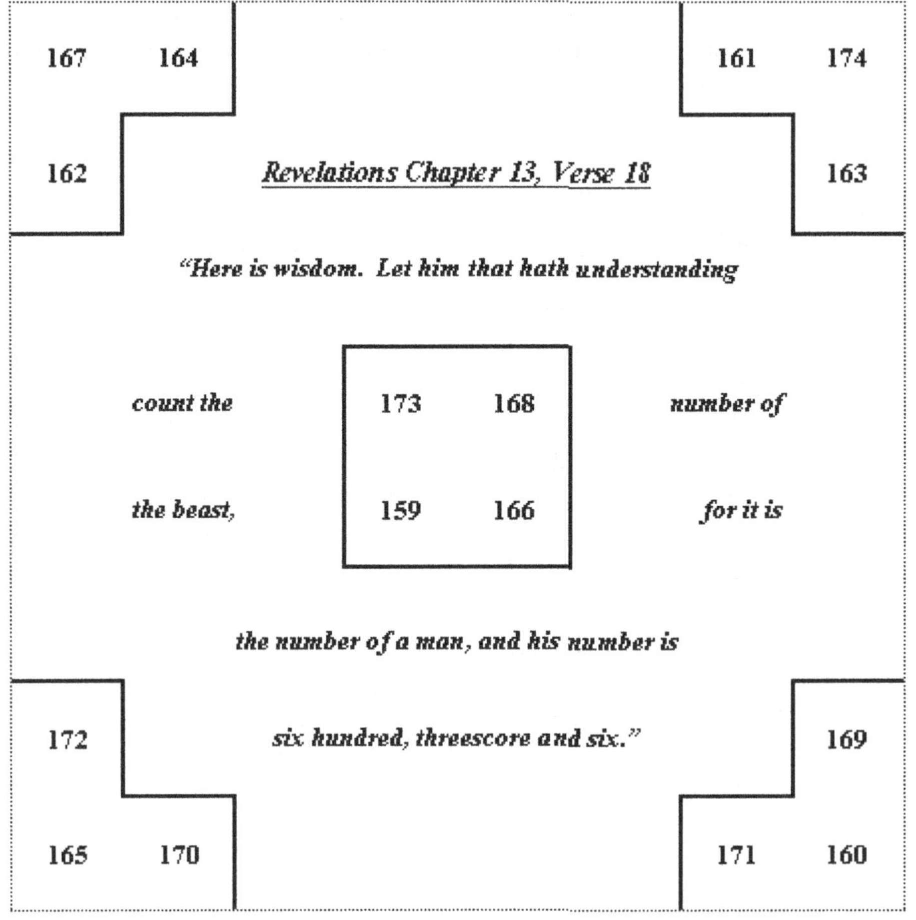

When the "Crossed Box Pleat" fold is made, a 4 x 4 magic square is produced, whose "magic total" is also 666:

A template for this Origami Magic Square is available in Appendix G, which you can cut out and photocopy. However, if you don't want to deface your copy of this book, you can download this template from http://www.MagicSquaresBook.com/.

This routine was originally developed for close-up, but, with larger props, it could also be presented in a cabaret situation.

Acknowledgements

This routine was inspired by a section in "The Magic Numbers Of Dr. Matrix" by Martin Gardner.

18.   **"Know .. Will .. Dare .. Keep Silent"**

This is a presentation for a lecture demonstration based upon magic squares, but also involving talismans, astrology and numerology. The title of the routine, which can be found in "Dynamic Mentalism" by Stanton Carlisle, is taken from the four precepts of "real" magi.

10.   References

The following lists include those books referenced in the above booklet. They are not intended to be exhaustive, but they **are** representative of the range of magical and mathematical literature that is available on this fascinating subject:

1.   Magic Squares

Andrews, W. S. (1960) *Magic Squares And Cubes*. New York, New York: Dover (and London: Constable & Company Limited)

> This is probably **the** definitive work on magic squares, how to construct them, the detailed mathematics behind them, and all of the many variations that exist. It is also a highly "technical" book, and of more interest to the serious mathematician than the average magician.

Anon. *Internet*

> This is a vast repository of electronically stored data, which contains information on practically every subject under the sun, including occasional "pages" about magic squares.

Anon. *The Complete Home Entertainer*. London: Odhams Press Limited

> This is quite an old book that I found on the local market. It contains a number of chapters on magic, and one on word games, one of which is the early Latin word square shown in Chapter 2, History, above.

Anon. (c. 1890) *National Encyclopaedia*.

> Volume IX of this Victorian encyclopaedia contains a small entry on magic squares, with an unusual example being shown of an 11 x 11 magic square that sums to a "magic total" of 24,156.

Barlow, F. (1951) *Mental Prodigies*. London: Hutchinson

> This book is primarily concerned with a study of mental prodigies (e.g. lightning calculators). However, a few chapters are devoted to the explanation of a few methods used by such people, including one brief chapter on magic squares.

Becker, L. (1986) *The Larry Becker Lecture*. London: Lewis Davenport Limited

> This video is a lecture on mentalism. It contains a number of unexplained items from Larry's show, including his famous "Russian Roulette" routine, a number of "lecture" items which are demonstrated and fully explained, and finally a number of marketed items, which are also demonstrated and explained. One of the "lecture" items is the one Larry calls "Déjà Vu", which utilises a magic square principle.

Benson, W. H. & Jacoby, O. (1976) *New Recreations With Magic Squares*. New York, New York: Dover (and London: Constable & Company Limited)

> This is a book that covers all orders of magic squares but is also oriented towards a mathematical analysis of magic squares rather than any magical presentations.

Benson, W. H. & Jacoby, O. (1981) *Magic Cubes*. New York, New York: Dover (and London: Constable & Company Limited)

> This book is entirely devoted to magic cubes, and follows on from many of the techniques published in their earlier book, "New Recreations With Magic Squares" (see below).

Blaisdell, F. (1978) *Magical Fun With Magic Squares*. Oakland, California: Magic Limited

> This book contains a number of innovative presentations of magic squares.

Brandreth, G. (1984) *Solo Games*. London: Pan Books

> This book contains a number of puzzles and games for one person. One of the puzzles involves creating a 6 x 6 magic square using 18 dominoes from a double-six set.

Britland, D. (2002) *The Mind & Magic Of David Berglas*. Burbank, California: Hahne Publications

> This is the long-awaited book that is more like a biography of David Berglas and his amazing magic than a typical magic book. It sold out of its 1,000 copies almost immediately. The book contains two excellent Magic Square routines.

Carlisle, S. (1979) *Dynamic Mentalism*. Bideford, Devon: The Supreme Magic Company

> This book, by one of the most well-known magicians in the field of mentalism, contains a lecture demonstration based upon magic squares, as well as a chapter on memory.

Clarke, J. (1969) *Number Relationships 3 - The Magic Square*. Leeds: E. J. Arnold & Son Limited

> This book is one of a series of "programmed learning" books for schoolchildren of at least ten years, and uses a 3 x 3 magic square to demonstrate the relationships between different numbers.

Corinda, A. (1968) *13 Steps To Mentalism.* New York, New York: Louis Tannen

> This is another classic book on mentalism, which has one lesson (Step 3, largely written by David Berglas) devoted to 5 x 5 magic squares, mnemonics and mental systems. For further information, the reader is referred to "Mental Prodigies" by Fred Barlow.

Dexter, W. (1974) *Feature Magic For Mentalists.* Bideford, Devon: The Supreme Magic Company

> This is a very useful book on mentalism, with one chapter which is devoted to mathematical magic. One part of this chapter is devoted to magic squares.

Encyclopaedia Britannica Inc. (1993) *Encyclopaedia Britannica ®.* Chicago, Illinois: Encyclopaedia Britannica Inc.

> This well-known encyclopaedia, which contains entries on thousands of subjects, has a short entry on magic squares (in Volume 7 of the 15th Edition of the Micropaedia Ready-Reference).

Falkener, E. (1961) *Games Ancient And Oriental And How To Play Them.* New York, New York: Dover

> This book, originally published in 1892, contains details of ancient Egyptian games (e.g. tau, senat, ham), chess, draughts, backgammon, and also magic squares and Knight's Tours.

Fulves, K. (1983) *Self-Working Number Magic.* New York, New York: Dover (and London: Constable & Company Limited)

> This is one of the excellent "Self-Working" series by this renowned author. This particular one contains a number of effects using the magic square principle.

Gardner, M. (1956) *Mathematics Magic And Mystery.* New York, New York: Dover (and London: Constable & Company Limited)

> This is a similar book to "Mathematical Magic Show" (see above). The book is full of tricks, games and puzzles, all of which have a mathematical nature.

Gardner, M. (1965) *Mathematical Puzzles And Diversions.* Harmondsworth, Middlesex: Pelican Books

> This book is entirely full of puzzles of a mathematical nature, but there is one chapter devoted to magic with a matrix (as used in the "Numorator" effect described in Chapter 9, Presentation / Routines, above).

Gardner, M. (1966) *More Mathematical Puzzles And Diversions*. Harmondsworth, Middlesex: Pelican Books

This is another book in the "leisure mathematics" series, which is full of tricks, games and puzzles, all of which have a mathematical nature. Chapter 12 is devoted to magic squares.

Gardner, M. (1978) *Mathematical Carnival*. Harmondsworth, Middlesex: Pelican Books

This is one of many books by this prolific author, whose interests include both magic and mathematics. There is one chapter devoted to magic stars and polyhedrons.

Gardner, M. (1985) *Mathematical Magic Show*. London: Penguin Books

This book is entirely devoted to mathematical magic, as its title suggests, with various effects using the magic square principle.

Gardner, M. (1985) *The Magic Numbers Of Dr. Matrix*. Buffalo, New York: Prometheus Books

This is a book which is primarily concerned with numerology and number theory, which mentions magic squares in several places.

Heath, R. V. (1953) *Mathemagic*. New York, New York: Dover (and London: Constable & Company Limited)

First written in 1933, this is a classic of mathematical magic. As far as magic squares are concerned, there is an entire chapter devoted to the creation and transposition of magic squares, and also other related formations (e.g. magic triangles).

Hilliard, J. N. (1994) *Greater Magic*. Kaufman & Greenberg

This is one of the classic magic books, recently republished, covering practically every subject possible. One chapter is devoted to the creation and presentation of magic squares.

I.B.M. *The Linking Ring*. St. Louis, Missouri: International Brotherhood Of Magicians

The October 1985 edition contains details of a force of the number 15, by Max Maven.

The July 1995 edition contains a routine by Gary Ward that uses the Max Maven force.

Johnson, R. (1995) *Unique*. Oadby, Leicestershire: Roy Johnson

This is one of Roy Johnson's set of lecture notes, which contains a book test, called "To Hell And Back", using a magic square principle.

Kane, P. (1989) *Kane At Table*. Peter Kane

> This is a set of lecture notes in which one of the explained effects has a magic square theme.

Kordemsky, B. (1975) *The Moscow Puzzles*. Harmondsworth, Middlesex: Pelican Books

> This is a book full of mathematical puzzles, but there is a small section (problems 331 - 339) on magic squares. It covers, briefly, the history of them, how to create them, and a few other interesting facts and problems.

Kraitchik, M. (1943) *Mathematical Recreations*. London: Allen & Unwin

> This is a book on general recreational mathematics, but it does include a section on magic squares. The book is quite detailed in nature, but is oriented towards a mathematical analysis of its contents rather than any magical presentations.

Lorayne, H. (1977) *The Magic Book*. London: W. H. Allen

> This is a magic book written for the general public as a beginners' book. One of the chapters covers number magic, with a method of creating an "instant" magic square included.

Meyer, O. (1961) *The Amazing Magic Square & Master Memory Demonstration*. Cheyenne, Wyoming: Orville Meyer

> This relatively short booklet contains one of the best routines to be based on magic squares, although it may be hard to find nowadays.

Moran, J. (1982) *The Wonders Of Magic Squares*. New York, New York: Random House, Inc.

> This book, written by a self-confessed "amateur recreational mathematician", is full of many different ways of creating magic squares.

Murray, B. (1989) *Paul Daniels' Adult Magic*. London: Michael O'Mara Books Limited

> This book is relatively well-known, and contains a variety of reasonably simple magic effects. There are a few pages on the subject of magic squares.

Raven, A. (1974) *The Necromantic Grimoire Of Augustus Rupp*. Waldwick, New Jersey: Bob Lynn

> This book, which is a limited edition, was specially printed to perform various magic effects. It "reproduces" a book of magic spells, symbols and other information, but includes two magic squares of the type shown in Chapter 2, History, of this booklet.

Riding, J. (1994) *Close Up Club.* St. Annes, Lancashire: Joe Riding

> This is a twelve-part course in close-up magic. Joe's new effect, "Computer Error", which is a routine for presenting a magic square, is included in the final lesson, but will also be published separately.

Simon, W. (1965) *Mathematical Magic.* London: Allen & Unwin

> This book is entirely devoted to mathematical magic, as its title suggests. However, there is one chapter on the subject of magic squares, including both how to create them and some ideas for using and presenting them.

Singer, J. (1994) *Microsoft Encarta* ®. Microsoft Corporation

> This is an encyclopaedia, stored on a CD-ROM, for use on a multimedia personal computer, which contains one entry on magic squares.

Smith, A. (1992) *Cards On Call.* Rotherham: Al Smith

> This book is not particularly related to mental magic or magic squares, as its title suggests. However, the routine described above, "Numorator", is an interesting addition to a performer's repertoire of magic square routines.

Tahan, M. (1994) *The Man Who Counted.* Edinburgh: Canongate Press

> This is a book that explores various aspects of mathematics via a series of stories told about a mathematician called Beremiz Samir. There are a couple of pages that describe, simply, magic squares.

Tucker, S. (1992) *Card Bored?*. Great Malvern, Worcestershire: Stephen Tucker

> This is a book of various close-up routines, mainly using playing cards. However, there are also three routines ("The Square Comes Full Circle", "The Almost Magic Square" and "Sweet Sixteen") that involve magic squares.

Van Delft, P. & Boterman, J. (1978) *Creative Puzzles Of The World.* London: Cassell Ltd.

> This book contains an excellent collection of puzzles from all over the world, as its title implies. There is a small but interesting and varied section on magic squares.

Wells, D. (1986) *The Penguin Dictionary Of Curious And Interesting Numbers.* London: Penguin Books

> Whilst not a magic book, this does contain some fascinating mathematical information, some of which can be used to enhance the presentation of a magic square demonstration.

2.   **Mnemonics**

Aronson, S. (1994) *Bound To Please*. Chicago, Illinois: Simon Aronson

Simon Aronson has created an innovative range of effects using a memorised deck. This book, which is a compilation of previously published books, contains a chapter called "A Stack To Remember", which provides full details of Simon Aronson's unique stacked deck, which may be used for various effects including poker deals, and also the method by which it may be memorised.

Berglas, D. & Playfair, G. L. (1988) *A Question Of Memory*. London: Jonathan Cape Limited

This book is entirely devoted to memory and mnemonic systems and draws extensively upon David's extensive knowledge of this subject. Of particular interest are the examples taken from one of his memory training seminars, where the difference between "good" associations and "bad" associations is made clear.

Buzan, A. (1971) *Speed Memory*. London: Sphere Books

This book contains numerous mnemonic systems, including the peg system.

Carlisle, S. (1979) *Dynamic Mentalism*. Bideford, Devon: The Supreme Magic Company

This book, by one of the most well-known magicians in the field of mentalism, contains a lecture demonstration based upon magic squares, as well as a chapter on memory.

Furst, B. (1977) *You Can Remember*. Marple, Cheshire: Memory & Concentration Studies

This is a twelve-part memory training course, which provides full details of all of the main mnemonic techniques. In addition, Session 10 is devoted to memorising playing cards.

Hancock, J. (1995) *Jonathan Hancock's Mindpower System*. London: Hodder & Stoughton

This is a relatively new book on memory training techniques by the 1994 World Memory Champion.

Harbin, R. (1968) *Instant Memory*. London: Corgi Mini-Books (Transworld Publishers Limited)

This small, pocket sized book was my first introduction to mnemonics. Whilst not a particularly deep book as far as they go, it does contain the basic techniques required for memorising lists of objects using both the link and peg methods.

Joseph, E. *Memory Of The Mind.* Bideford, Devon: The Supreme Magic Company

> This book contains both routines and systems for mental effects and, more importantly, memorising a pack of playing cards.

Lorayne, H. & Lucas, J. (1976) *The Memory Book.* London: W. H. Allen & Company Limited

> This book is, as its title suggests, entirely about mnemonic systems. Harry Lorayne is, of course, one of the best known practitioners of the mnemonic art. The book tells you all you need to know about memorising anything, including playing cards.

Mirsky, N. (1994) *The Unforgettable Memory Book.* London: Penguin Books (BBC Books)

> This book on mnemonics is the one that accompanies the BBC series, presented by Greg Proops.

O'Brien, D. (1993) *How To Develop A Perfect Memory.* London: Pavilion Books Limited

> This book is entirely devoted to mnemonic systems, including memorising an entire pack of playing cards. However, the systems used by Dominic O'Brien, who has won the World Memory Championship several times, are mainly different to those used by Harry Lorayne: some are old and some are Dominic's own, new, personal ideas. (The old system is that first used by the ancient Greek and Roman orators. They used a method which associated ideas to known places or routes with which they were familiar. This has given rise to the now popular phrase "in the first place".)

3.   **Other**

Mulatinho, P. (1995) *Origami.* London: Grange Books plc

> This book contains a number of interesting origami figures, including the "Crossed Box Pleat" used in my "Origami Magic Square" and "The Number Of The Beast" routines, above.

Total Combinations Of Four Squares That Add Up To A Selected Total
(Complex Formula)

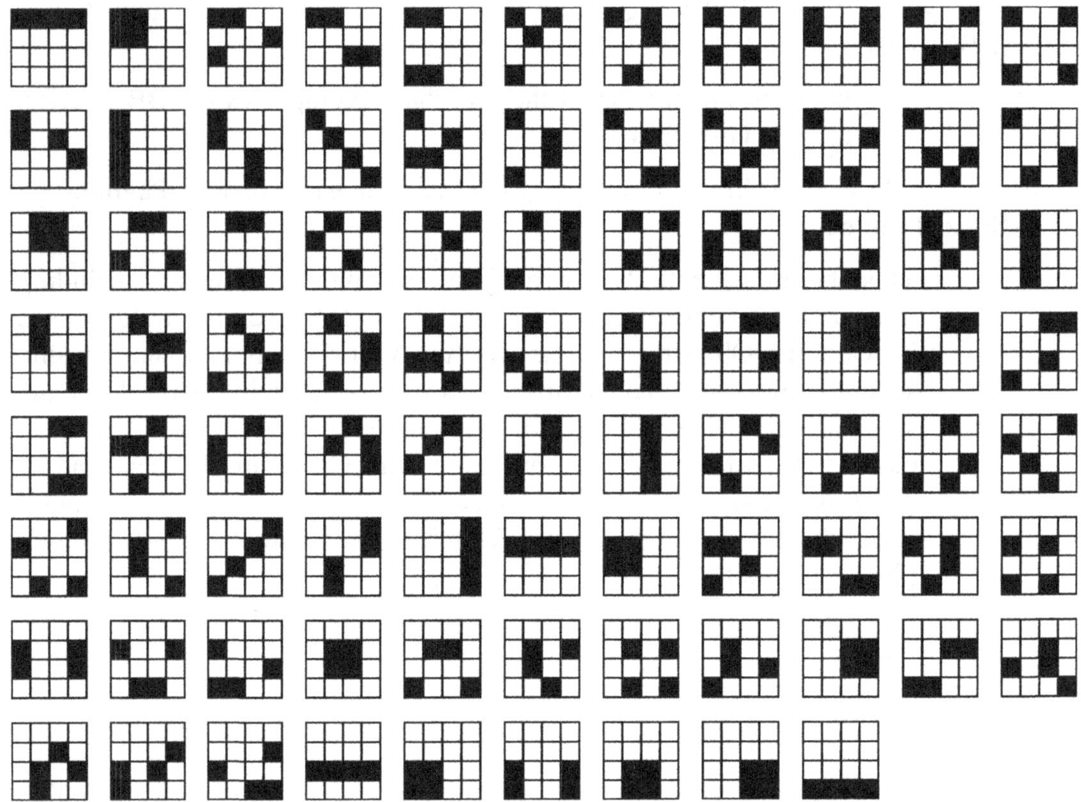

Total Combinations Of Four Squares That Add Up To A Selected Total
(Complex Formula)

The following **86** combinations of four squares, and those shown on the facing page, add up to the "magic total" in a magic square whose magic total has a remainder of **zero** once 34 has been subtracted from that magic total and the result of that subtraction has been divided by four. (Note that those combinations shown in bold print are those that are common to all 4 x 4 magic squares.)

ABCD	**BEGI**	**DHJN**
ABEF	**BELO**	**DHLP**
ABHI	**BFHK**	**EFGH**
ABKL	**BFJN**	EFIJ
ABMN	**BFLP**	**EFKM**
ACFM	**BGHO**	EFOP
ACGN	BGLM	EGKN
ACIK	BHLN	**EGMO**
ADEH	BIJO	**EHIL**
ADJK	**BINP**	**EHNO**
ADMP	**BKMO**	ELMN
AEGL	**CDEL**	**FGJK**
AEIM	**CDGH**	**FGMP**
AEKO	CDIJ	FHJO
AFKP	CDKM	**FHNP**
AGIJ	CDOP	**FJLM**
AGKM	**CEFN**	GHKL
AGOP	CEIO	GHMN
AHKN	**CFHL**	**GIKP**
AHMO	CFIP	GJLN
AJLO	**CGIM**	**HIKM**
ALNP	**CGKO**	HIOP
BCFG	**CHIN**	**IJKL**
BCIL	CKLN	**IJMN**
BCNO	**CLMO**	**ILMP**
BDEK	DEJO	**JKNO**
BDGP	**DENP**	**KLOP**
BDHM	DFJP	**MNOP**
BDJL	**DGJM**	

Total Combinations Of Four Squares That Add Up To A Selected Total
(Complex Formula)

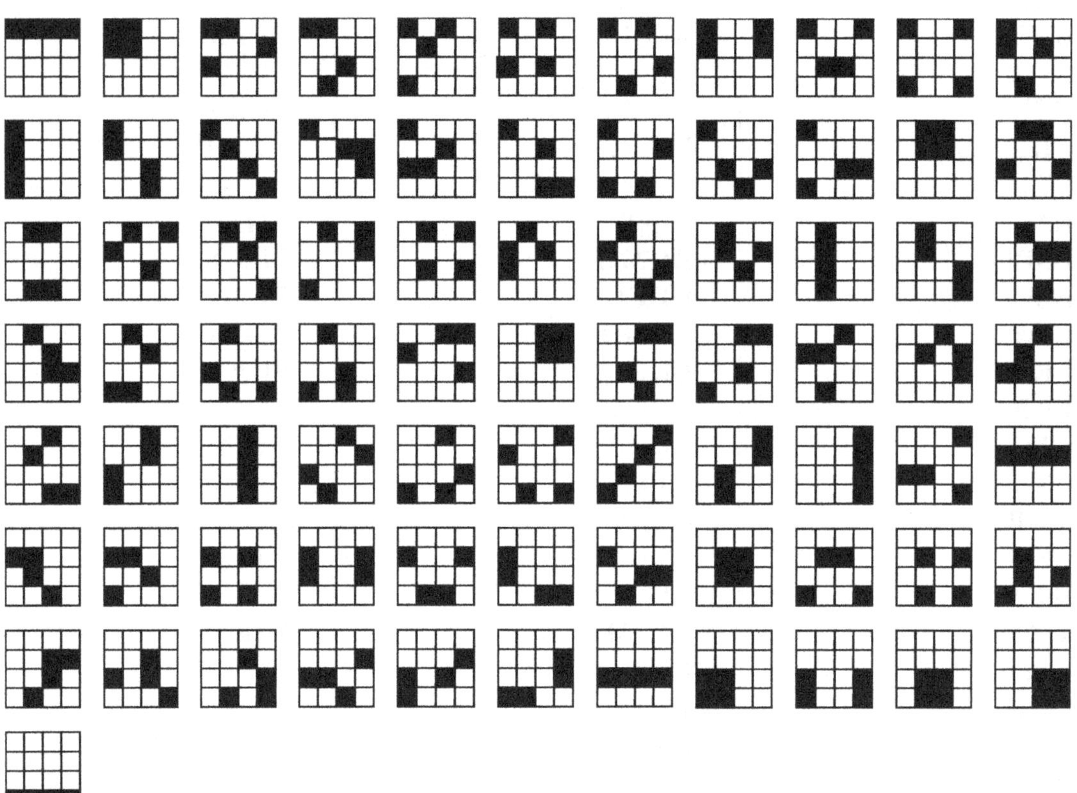

Total Combinations Of Four Squares That Add Up To A Selected Total
(Complex Formula)

The following **78** combinations of four squares, and those shown on the facing page, add up to the "magic total" in a magic square whose magic total has a remainder of **one** once 34 has been subtracted from that magic total and the result of that subtraction has been divided by four. (Note that those combinations shown in bold print are those that are common to all 4 x 4 magic squares.)

ABCD	**BDJL**	**DHLP**
ABEF	**BEGI**	DIJP
ABHI	**BELO**	**EFGH**
ABKN	**BFHK**	EFJO
ACFM	**BFJN**	**EFKM**
ACIK	**BFLP**	**EGMO**
ACLN	**BGHO**	**EHIL**
ADEH	BGKL	**EHNO**
ADJK	BGMN	EIOP
ADMP	**BINP**	EKLN
AEGN	**BKMO**	**FGJK**
AEIM	**CDEL**	**FGMP**
AEKO	**CDGH**	**FHNP**
AFKP	CDJO	**FJLM**
AGHL	**CDKM**	GHKN
AGIJ	**CEFN**	**GIKP**
AGOP	**CFHL**	GLNP
AHMO	CFIJ	HIJO
AJLO	CFOP	**HIKM**
AKLM	**CGIM**	HLMN
BCFG	**CGKO**	**IJKL**
BCIL	**CHIN**	**IJMN**
BCNO	**CLMO**	**ILMP**
BDEK	**DENP**	**JKNO**
BDGP	**DGJM**	**KLOP**
BDHM	**DHJN**	**MNOP**

Total Combinations Of Four Squares That Add Up To A Selected Total
(Complex Formula)

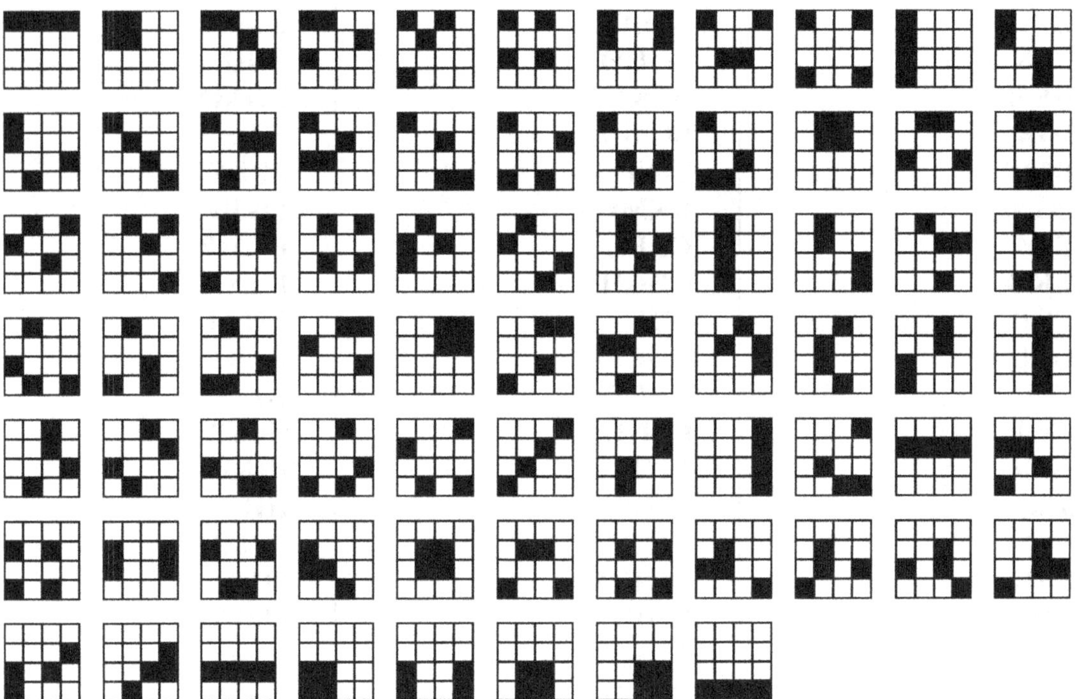

Total Combinations Of Four Squares That Add Up To A Selected Total
(Complex Formula)

The following **74** combinations of four squares, and those shown on the facing page, add up to the "magic total" in a magic square whose magic total has a remainder of **two** once 34 has been subtracted from that magic total and the result of that subtraction has been divided by four. (Note that those combinations shown in bold print are those that are common to all 4 x 4 magic squares.)

ABCD	**BDJL**	**DHJN**
ABEF	**BEGI**	**DHLP**
ABGL	**BELO**	DJOP
ABHI	**BFHK**	**EFGH**
ACFM	**BFJN**	**EFKM**
ACIK	**BFLP**	**EGMO**
ADEH	**BGHO**	**EHIL**
ADJK	BGKN	**EHNO**
ADMP	**BINP**	EIJO
AEIM	**BKMO**	**FGJK**
AEKO	BLMN	**FGMP**
AELN	**CDEL**	**FHNP**
AFKP	**CDGH**	FIJP
AGHN	**CDKM**	**FJLM**
AGIJ	**CEFN**	**GIKP**
AGOP	**CFHL**	GKLM
AHMO	CFJO	**HIKM**
AJLO	**CGIM**	HKLN
AKMN	**CGKO**	**IJKL**
BCFG	CGLN	**IJMN**
BCIL	**CHIN**	**ILMP**
BCNO	CIOP	**JKNO**
BDEK	**CLMO**	**KLOP**
BDGP	**DENP**	**MNOP**
BDHM	**DGJM**	

Total Combinations Of Four Squares That Add Up To A Selected Total
(Complex Formula)

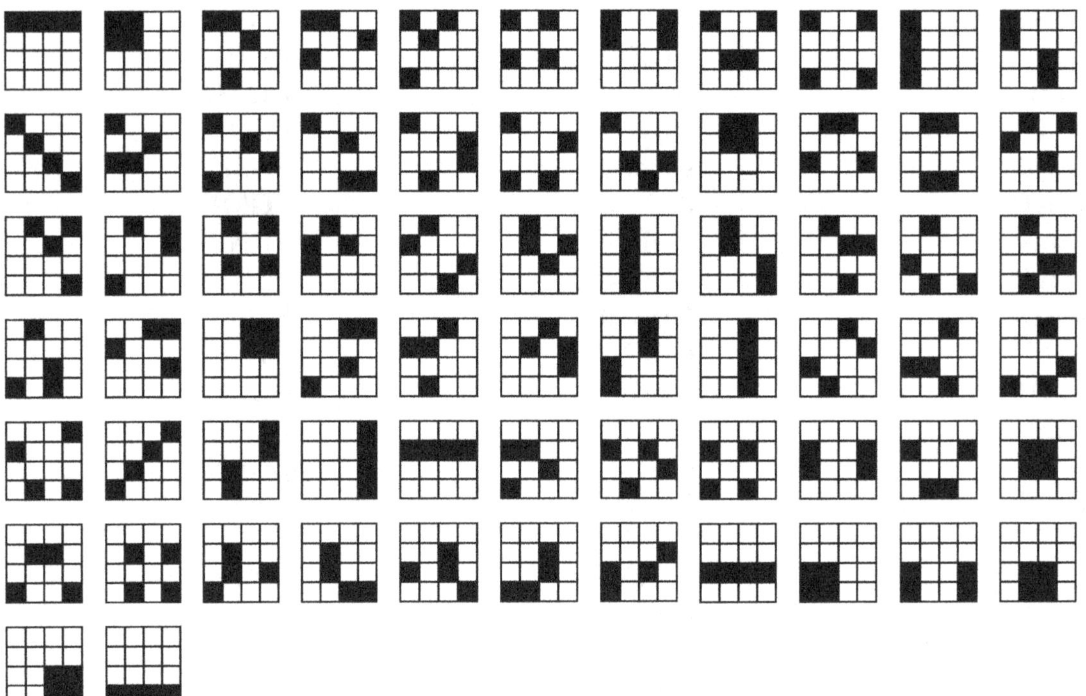

Total Combinations Of Four Squares That Add Up To A Selected Total
(Complex Formula)

The following **68** combinations of four squares, and those shown on the facing page, add up to the "magic total" in a magic square whose magic total has a remainder of **three** once 34 has been subtracted from that magic total and the result of that subtraction has been divided by four. (Note that those combinations shown in bold print are those that are common to all 4 x 4 magic squares.)

ABCD	**CEFN**
ABEF	**CFHL**
ABGN	**CGIM**
ABHI	**CGKO**
ACFM	**CHIN**
ACIK	CIJO
ADEH	**CLMO**
ADJK	**DENP**
ADMP	**DGJM**
AEIM	**DHJN**
AEKO	**DHLP**
AFKP	**EFGH**
AGIJ	**EFKM**
AGLM	EGLN
AGOP	**EGMO**
AHLN	**EHIL**
AHMO	**EHNO**
AJLO	**FGJK**
BCFG	**FGMP**
BCIL	**FHNP**
BCNO	**FJLM**
BDEK	FJOP
BDGP	**GIKP**
BDHM	GKMN
BDJL	**HIKM**
BEGI	**IJKL**
BELO	**IJMN**
BFHK	**ILMP**
BFJN	**JKNO**
BFLP	**KLOP**
BGHO	**MNOP**
BINP	
BKLN	
BKMO	
CDEL	
CDGH	
CDKM	

This page has been intentionally left blank

Summary Of Number Of Combinations Of Four Squares That Add Up To A Selected Total (Simple Formula)

The following table shows the number of combinations of four squares that will sum to a specified "magic total" in a 4 x 4 magic square when the "simple" formula of creating the magic square is used:

"Magic Total"	Number Of Combinations
34	86
35	78
36	74
37	68
38	66
39	63
40	62
41	61
42	61
All Others	60

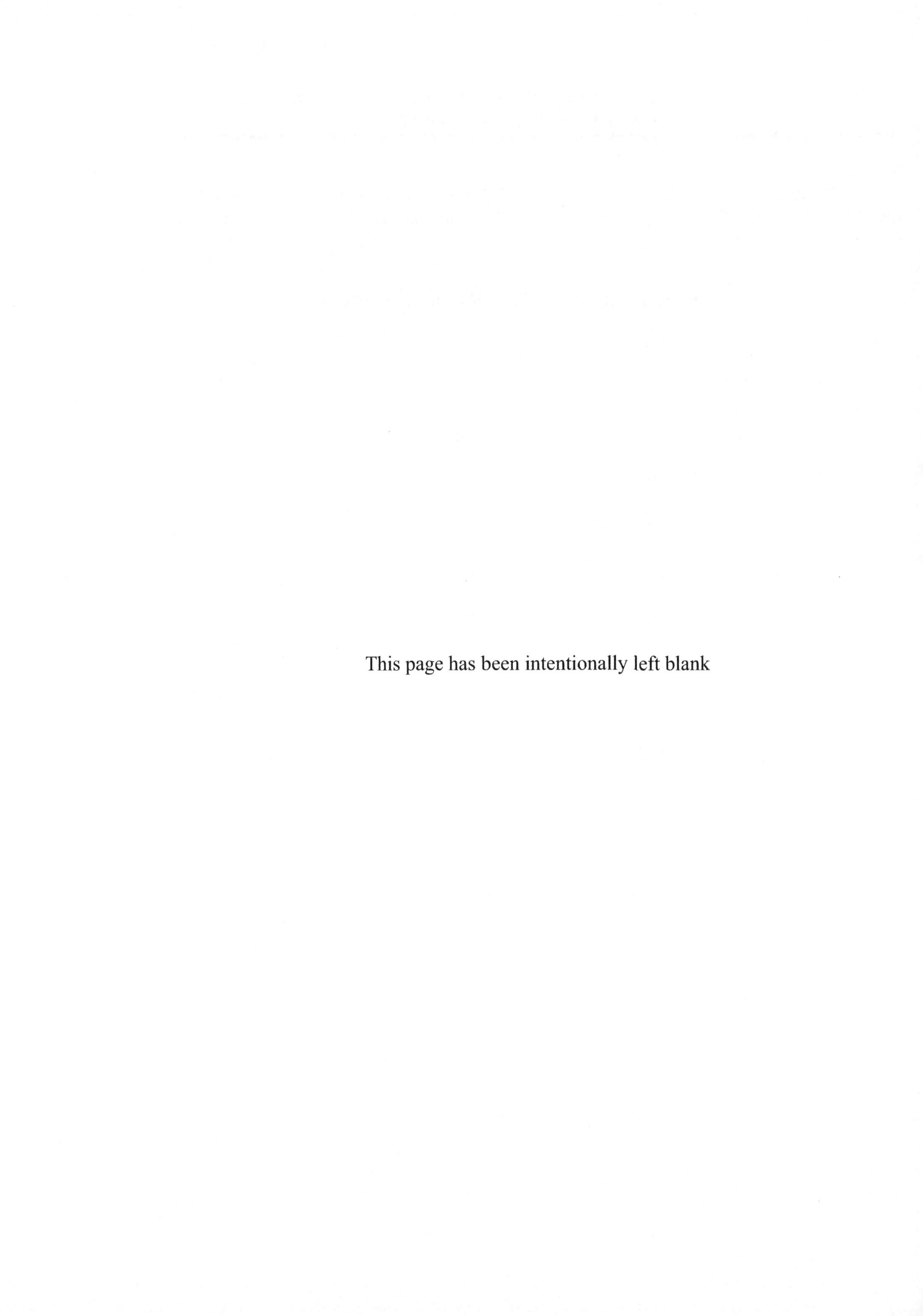

This page has been intentionally left blank

Total Combinations Of Five Squares That Add Up To A Selected Total
(Complex Formula)

The following **1,394** combinations of five squares add up to the magic total in a magic square whose "magic total" has a remainder of **zero** once 65 has been subtracted from that magic total and the result of that subtraction has been divided by five. (Note that those combinations shown in bold print are those that are common to all 5 x 5 magic squares.)

ABCDE	**ABSTW**	ADLMX	**AFMSU**	**AJKRY**	**BCFNO**
ABCFW	**ACDFU**	**ADLUV**	**AFNPW**	**AJKTW**	**BCGHT**
ABCGV	**ACDGY**	**ADLWY**	**AFNRU**	**AJLMR**	BCGIS
ABCIY	**ACDIW**	ADMOU	AFNTX	**AJLTV**	BCGKL
ABCJM	**ACDJV**	**ADNUY**	AFOPV	**AJMNP**	BCGPU
ABCKR	**ACDKP**	ADNVX	**AFORY**	AJMOT	**BCGRX**
ABCNT	**ACDLT**	ADOPR	**AFOTW**	**AJMSV**	**BCHIR**
ABDGO	**ACDMS**	ADOVW	AFPQT	**AJNRV**	**BCHKV**
ABDHY	**ACDNR**	ADPQV	**AFPRS**	**AJNTY**	**BCHMN**
ABDJL	**ACEMR**	**ADPSY**	AFQVY	**AKLMW**	**BCIKO**
ABDKQ	**ACETV**	ADQRY	**AFSVW**	**AKLPT**	BCILN
ABDNS	**ACFGR**	ADQTW	AGIMX	**AKLVY**	**BCIPX**
ABDWX	**ACFIP**	**ADRSW**	**AGIUV**	**AKMNU**	BCJKN
ABEFU	**ACFJT**	**ADSTU**	**AGIWY**	AKMOY	BCJPW
ABEGY	**ACFKY**	**AEFGP**	**AGJMW**	**AKMPS**	BCJRU
ABEIW	**ACFLM**	**AEFJR**	**AGJPT**	AKMQR	**BCJTX**
ABEJV	**ACFNV**	**AEFKW**	**AGJVY**	**AKNPR**	BCKUW
ABEKP	**ACGIT**	**AEFLV**	**AGKPY**	**AKNVW**	**BCKXY**
ABELT	**ACGKM**	AEFMO	**AGKRW**	AKORT	**BCLMX**
ABEMS	**ACIKV**	**AEFNY**	**AGKTU**	AKQTV	**BCLUV**
ABENR	**ACIMN**	**AEFST**	**AGLMP**	**AKRSV**	**BCLWY**
ABFGS	**ACMUV**	**AEGIR**	**AGLRV**	**AKSTY**	**BCMOU**
ABFHR	**ACMWY**	**AEGKV**	**AGLTY**	**ALMNY**	BCNUY
ABFIQ	**ACPRV**	**AEGMN**	AGMOR	**ALMST**	BCNVX
ABFKO	**ACPTY**	**AEIJT**	AGMQV	**ALNRT**	**BCOPR**
ABFLN	**ACRTW**	**AEIKY**	**AGMSY**	AMNOV	**BCOVW**
ABFPX	ADEFX	**AEILM**	**AGNPV**	AMNQT	**BCPQV**
ABGIJ	**ADEGW**	**AEINV**	**AGNRY**	**AMNRS**	**BCPSY**
ABGKN	**ADEIU**	**AEJKM**	**AGNTW**	**AMPUW**	**BCQRY**
ABGPW	**ADEJY**	**AEKNT**	AGOTV	AMPXY	**BCQTW**
ABGRU	**ADELR**	**AEMUY**	**AGRST**	AMRWX	**BCRSW**
ABGTX	ADEMQ	AEMVX	AHMPV	AMTUX	**BCSTU**
ABHIT	**ADENP**	**AEPRY**	AHMRY	**ANSTV**	**BDEHU**
ABHKM	ADEOT	**AEPTW**	AHMTW	APRTX	**BDELQ**
ABIKL	**ADESV**	**AERTU**	AHTVY	**APUVY**	**BDEOS**
ABIPU	ADFGQ	**AEVWY**	**AIJMU**	**ARUVW**	**BDGHS**
ABIRX	ADFHP	AFGHM	**AIJPR**	ARVXY	BDGQX
ABJPY	**ADFJS**	**AFGIL**	**AIJVW**	**ATUWY**	BDHIQ
ABJRW	ADFNO	**AFGJK**	**AIKPW**	ATVWX	BDHKO
ABJTU	ADGHT	**AFGUY**	**AIKRU**	**BCDHW**	**BDHLN**
ABKUY	**ADGIS**	AFGVX	AIKTX	**BCDJO**	BDHPX
ABKVX	**ADGKL**	AFHIV	**AILPV**	BCDLS	**BDJQU**
ABLMU	**ADGPU**	**AFHKT**	**AILRY**	**BCDNQ**	**BDJSX**
ABLPR	ADGRX	**AFIJN**	**AILTW**	**BCDUX**	**BDLOU**
ABLVW	ADHIR	**AFIKS**	AIMOP	**BCEFX**	BDNOX
ABMNX	ADHKV	**AFIUW**	AIMQY	**BCEGW**	BDOPQ
ABMOW	ADHMN	AFIXY	**AIMSW**	BCEIU	**BDQSW**
ABMPQ	ADIKO	AFJMX	**AINPY**	BCEJY	**BEFJQ**
ABNUV	**ADILN**	**AFJUV**	AINRW	BCELR	**BEFLO**
ABNWY	ADIPX	**AFJWY**	**AINTU**	**BCEMQ**	**BEGHR**
ABOPT	**ADJKN**	AFKPU	AIORV	BCENP	**BEGIQ**
ABOVY	**ADJPW**	AFKRX	AIOTY	**BCEOT**	**BEGKO**
ABPSV	**ADJRU**	**AFLPY**	**AIPST**	BCESV	BEGLN
ABQRV	**ADJTX**	**AFLRW**	AIQRT	**BCFGQ**	**BEGPX**
ABQTY	**ADKUW**	**AFLTU**	**AISVY**	**BCFHP**	**BEHIP**
ABRSY	ADKXY	AFMQW	**AJKPV**	BCFJS	**BEHJT**

Total Combinations Of Five Squares That Add Up To A Selected Total
(Complex Formula)

BEHKY	**BHIUY**	**BLRWX**	**CEFQT**	CGKRU	**CKOPT**
BEHLM	BHIVX	**BLTUX**	CEFRS	**CGKTX**	**CKOVY**
BEHNV	**BHJMU**	**BMNQS**	CEGIP	CGLPV	CKPSV
BEIJS	**BHJPR**	BMQWX	CEGJT	CGLRY	**CKQRV**
BEINO	**BHJVW**	**BMSUX**	CEGKY	CGLTW	**CKQTY**
BEJKL	**BHKPW**	**BNOST**	CEGLM	**CGMOP**	CKRSY
BEJPU	**BHKRU**	**BNPWX**	CEGNV	**CGMQY**	CKSTW
BEJRX	BHKTX	**BNRUX**	**CEHMV**	CGMSW	CLMNW
BEKNS	**BHLPV**	**BOPUY**	CEIJR	CGNPY	**CLMOV**
BEKWX	**BHLRY**	BOPVX	CEIKW	CGNRW	**CLMQT**
BELUY	**BHLTW**	**BORUW**	CEILV	CGNTU	CLMRS
BELVX	BHMOP	BORXY	**CEIMO**	**CGORV**	CLNPT
BEMOX	BHMQY	BOTWX	CEINY	**CGOTY**	CLNVY
BENUW	**BHMSW**	**BPQRU**	CEIST	CGPST	CLSTV
BENXY	**BHNPY**	BPQTX	CEJKV	**CGQRT**	**CMNOY**
BEOUV	**BHNRW**	**BPRSX**	CEJMN	CGSVY	CMNPS
BEOWY	**BHNTU**	**BQUVW**	CEKLT	**CHIKT**	**CMNQR**
BEPQY	BHORV	BQVXY	CEKMS	**CHMPY**	**CMOST**
BEPSW	BHOTY	BSUWY	CEKNR	**CHMRW**	**CMPWX**
BEQRW	**BHPST**	**BSVWX**	CEMUW	**CHMTU**	**CMRUX**
BEQTU	BHQRT	**CDEFH**	**CEMXY**	**CHPRT**	**CNORT**
BERSU	**BHSVY**	CDEGU	CEPRW	**CHRVY**	**CNQTV**
BESTX	BIJLU	**CDEIX**	CEPTU	**CHTVW**	CNRSV
BFGHL	**BIJNX**	CDEJW	**CERTX**	**CIJMX**	CNSTY
BFGOX	**BIJOW**	CDELP	CEUVY	CIJUV	CPUVW
BFHIO	**BIJPQ**	**CDEOR**	**CFGHV**	CIJWY	**CPVXY**
BFHJN	**BIKQU**	**CDEQV**	**CFGIO**	CIKPU	CRUWY
BFHKS	**BIKSX**	CDESY	CFGJN	**CIKRX**	**CRVWX**
BFHUW	**BILOP**	**CDFJQ**	CFGKS	CILPY	**CTUVX**
BFHXY	**BILQY**	**CDFLO**	CFGUW	CILRW	**CTWXY**
BFJLX	BILSW	**CDGHR**	**CFGXY**	CILTU	**DEFQS**
BFJOU	**BINQW**	**CDGIQ**	**CFHIY**	**CIMQW**	**DEGHP**
BFKQX	BINSU	**CDGKO**	**CFHJM**	CIMSU	DEGJS
BFLQW	BIOQV	CDGLN	**CFHKR**	CINPW	**DEGNO**
BFLSU	**BIOSY**	**CDGPX**	**CFHNT**	CINRU	**DEHJR**
BFNQU	**BIQRS**	**CDHIP**	CFIJL	**CINTX**	**DEHKW**
BFNSX	**BIUWX**	**CDHJT**	**CFIKQ**	**CIOPV**	**DEHLV**
BFOQY	**BJKOP**	**CDHKY**	CFINS	**CIORY**	DEHMO
BFOSW	**BJKQY**	**CDHLM**	**CFIWX**	**CIOTW**	**DEHNY**
BFPQS	BJKSW	**CDHNV**	CFJUY	**CIPQT**	**DEHST**
BGHIN	**BJLMQ**	CDIJS	**CFJVX**	CIPRS	**DEIJQ**
BGHMX	BJLNP	**CDINO**	CFKLN	**CIQVY**	**DEILO**
BGHUV	**BJLOT**	CDJKL	**CFKPX**	CISVW	**DEJKO**
BGHWY	BJLSV	CDJPU	CFLPW	CJKPY	DEJLN
BGILX	**BJMOS**	**CDJRX**	CFLRU	CJKRW	**DEJPX**
BGIOU	**BJNOR**	CDKNS	**CFLTX**	CJKTU	DEKLS
BGJKX	**BJNQV**	**CDKWX**	**CFMQU**	CJLMP	**DEKNQ**
BGJLW	BJNSY	CDLUY	**CFMSX**	CJLRV	**DEKUX**
BGJNU	**BJUVX**	**CDLVX**	CFNPU	CJLTY	DELUW
BGJOY	**BJWXY**	CDMOX	**CFNRX**	**CJMOR**	**DELXY**
BGJPS	BKLNU	CDNUW	**CFOPY**	**CJMQV**	**DENWX**
BGJQR	**BKLOY**	**CDNXY**	**CFORW**	CJMSY	**DEOUY**
BGKQW	BKLPS	**CDOUV**	**CFOTU**	CJNPV	DEOVX
BGKSU	**BKLQR**	**CDOWY**	**CFPQR**	CJNRY	**DEPQW**
BGLOR	**BKNOW**	**CDPQY**	**CFQVW**	CJNTW	DEPSU
BGLQV	**BKNPQ**	CDPSW	CFSUV	**CJOTV**	**DEQRU**
BGLSY	BKOQT	**CDQRW**	CFSWY	CJRST	DEQTX
BGMOQ	**BKORS**	**CDQTU**	**CGHIM**	CKLMU	**DERSX**
BGNOP	**BKPUX**	CDRSU	CGIJK	CKLPR	DFGHO
BGNQY	**BKQSV**	**CDSTX**	CGIUY	CKLVW	**DFHJL**
BGNSW	**BLNOV**	CEFJP	**CGIVX**	**CKMNX**	DFHKQ
BGOSV	**BLNQT**	CEFKU	CGJMU	**CKMOW**	**DFHNS**
BGQST	BLNRS	CEFLY	CGJPR	**CKMPQ**	DFHWX
BGUXY	**BLPUW**	CEFNW	CGJVW	CKNUV	DFJOX
BHIJK	**BLPXY**	**CEFOV**	CGKPW	CKNWY	**DFLQU**

Total Combinations Of Five Squares That Add Up To A Selected Total
(Complex Formula)

DFLSX	DJNQY	EGIJN	EJORV	FGNPQ	FLNSY
DFNQX	DJNSW	EGIKS	EJOTY	FGOQT	FLUVX
DFOQW	DJOSV	EGIUW	EJPST	FGORS	FLWXY
DFOSU	DJQST	EGIXY	EJQRT	FGPUX	FMNOQ
DGHIL	DJUXY	EGJMX	EJSVY	FGQSV	FMOUX
DGHJK	DKLNX	EGJUV	EKLMX	FHIKL	FNOSV
DGHUY	DKLOW	EGJWY	EKLUV	FHIPU	FNQST
DGHVX	DKLPQ	EGKPU	EKLWY	FHIRX	FNUXY
DGIOX	DKNOU	EGKRX	EKMOU	FHJPY	FOPRX
DGJLU	DKOPS	EGLPY	EKNUY	FHJRW	FOUWY
DGJNX	DKOQR	EGLRW	EKNVX	FHJTU	FOVWX
DGJOW	DKQSY	EGLTU	EKOPR	FHKUY	FPQUY
DGJPQ	DLMQS	EGMQW	EKOVW	FHKVX	FPQVX
DGKQU	DLNOY	EGMSU	EKPQV	FHLMU	FPSUW
DGKSX	DLNPS	EGNPW	EKPSY	FHLPR	FPSXY
DGLOP	DLNQR	EGNRU	EKQRY	FHLVW	FQRUW
DGLQY	DLOST	EGNTX	EKQTW	FHMNX	FQRXY
DGLSW	DLPWX	EGOPV	EKRSW	FHMOW	FQTWX
DGNQW	DLRUX	EGORY	EKSTU	FHMPQ	FRSWX
DGNSU	DMQUX	EGOTW	ELMNU	FHNUV	FSTUX
DGOQV	DNOQT	EGPQT	ELMOY	FHNWY	GHIPY
DGOSY	DNORS	EGPRS	ELMPS	FHOPT	GHIRW
DGQRS	DNPUX	EGQVY	ELMQR	FHOVY	GHITU
DGUWX	DNQSV	EGSVW	ELNPR	FHPSV	GHJMP
DHIJN	DOPUW	EHIJM	ELNVW	FHQRV	GHJRV
DHIKS	DOPXY	EHIKR	ELORT	FHQTY	GHJTY
DHIUW	DORWX	EHINT	ELQTV	FHRSY	GHKMU
DHIXY	DOTUX	EHKMN	ELRSV	FHSTW	GHKPR
DHJMX	DPQRX	EHMPW	ELSTY	FIJQW	GHKVW
DHJUV	DQUWY	EHMRU	EMNOW	FIJSU	GHLMY
DHJWY	DQVWX	EHMTX	EMNPQ	FIKOX	GHLRT
DHKPU	DSUVX	EHPVY	EMOQT	FILNX	GHMNW
DHKRX	DSWXY	EHRVW	EMORS	FILOW	GHMOV
DHLPY	EFGHY	EHTUV	EMPUX	FILPQ	GHMQT
DHLRW	EFGJL	EHTWY	EMQSV	FINOU	GHMRS
DHLTU	EFGKQ	EIJUY	ENOPT	FIOPS	GHNPT
DHMQW	EFGNS	EIJVX	ENOVY	FIOQR	GHNVY
DHMSU	EFGWX	EIKLN	ENPSV	FIQSY	GHSTV
DHNPW	EFHIW	EIKPX	ENQRV	FJKLU	GIJLP
DHNRU	EFHJV	EILPW	ENQTY	FJKNX	GIJOR
DHNTX	EFHKP	EILRU	ENRSY	FJKOW	GIJQV
DHOPV	EFHLT	EILTX	ENSTW	FJKPQ	GIJSY
DHORY	EFHMS	EIMQU	EOSTV	FJLNW	GIKLU
DHOTW	EFHNR	EIMSX	EPUWY	FJLOV	GIKNX
DHPQT	EFIJO	EINPU	EPVWX	FJLQT	GIKOW
DHPRS	EFILS	EINRX	ERUVX	FJLRS	GIKPQ
DHQVY	EFINQ	EIOPY	ERWXY	FJMQS	GILNW
DHSVW	EFIUX	EIORW	ETUXY	FJNOY	GILOV
DIJLX	EFJKS	EIOTU	FGHIJ	FJNPS	GILQT
DIJOU	EFJUW	EIPQR	FGHKN	FJNQR	GILRS
DIKQX	EFJXY	EIQVW	FGHPW	FJOST	GIMQS
DILQW	EFKNO	EISUV	FGHRU	FJPWX	GINOY
DILSU	EFLPU	EISWY	FGHTX	FJRUX	GINPS
DINQU	EFLRX	EJKPW	FGIQU	FKLOP	GINQR
DINSX	EFMQX	EJKRU	FGISX	FKLQY	GIOST
DIOQY	EFNPX	EJKTX	FGJOP	FKLSW	GIPWX
DIOSW	EFOPW	EJLPV	FGJQY	FKNQW	GIRUX
DIPQS	EFORU	EJLRY	FGJSW	FKNSU	GJKLY
DJKQW	EFOTX	EJLTW	FGKLX	FKOQV	GJKNW
DJKSU	EFQUV	EJMOP	FGKOU	FKOSY	GJKOV
DJLOR	EFQWY	EJMQY	FGLNU	FKQRS	GJKQT
DJLQV	EFSUY	EJMSW	FGLOY	FKUWX	GJKRS
DJLSY	EFSVX	EJNPY	FGLPS	FLMOS	GJLNV
DJMOQ	EGHIV	EJNRW	FGLQR	FLNOR	GJMNO
DJNOP	EGHKT	EJNTU	FGNOW	FLNQV	GJNST

Total Combinations Of Five Squares That Add Up To A Selected Total
(Complex Formula)

GJPUY	HIJTW	HRTXY	IQRVX	KLNST	LPQTY
GJPVX	HIKMX	IJKLW	IQTUW	KLPUY	LPRSY
GJRUW	HIKUV	IJKNU	IQTXY	KLPVX	LPSTW
GJRXY	HIKWY	IJKOY	IRSUW	KLRUW	LQRTW
GJTWX	HILMW	IJKPS	IRSXY	KLRXY	LRSTU
GKLMQ	HILPT	IJKQR	ISTWX	KLTWX	LSVWY
GKLNP	HILVY	IJLMO	JKLMS	KMOPX	MNOPU
GKLOT	HIMNU	IJLNY	JKLNR	KMQUW	MNORX
GKLSV	HIMOY	IJLST	JKMNQ	KMQXY	MNQUY
GKMOS	HIMPS	IJNOV	JKMUX	KMSWX	MNQVX
GKNOR	HIMQR	IJNQT	JKNOT	KNPUW	MNSUW
GKNQV	HINPR	IJNRS	JKNSV	KNPXY	MNSXY
GKNSY	HINVW	IJPUW	JKPRX	KNRWX	MOPQR
GKUVX	HIORT	IJPXY	JKUWY	KNTUX	MOQVW
GKWXY	HIQTV	IJRWX	JKVWX	KOPUV	MOSUV
GLMNS	HIRSV	IJTUX	JLMUW	KOPWY	MOSWY
GLMWX	HISTY	IKLOR	JLMXY	KORUY	MPQSY
GLPRU	HJKMW	IKLQV	JLPRW	KORVX	MQRSW
GLPTX	HJKPT	IKLSY	JLPTU	KOTUW	MQSTU
GLUVW	HJKVY	IKMOQ	JLRTX	KOTXY	NOPVW
GLVXY	HJLMV	IKNOP	JLUVY	KPQRW	NORUV
GMNUX	HJMNY	IKNQY	JMNWX	KPQTU	NORWY
GMOUW	HJMST	IKNSW	JMOUY	KPRSU	NOTUY
GMOXY	HJNRT	IKOSV	JMOVX	KPSTX	NOTVX
GMPQU	HKLMP	IKQST	JMPQW	KQRTX	NPQRY
GMPSX	HKLRV	IKUXY	JMPSU	KQUVY	NPQTW
GMQRX	HKLTY	ILMNQ	JMQRU	KSUVW	NPRSW
GNPRX	HKMOR	ILMUX	JMQTX	KSVXY	NPSTU
GNUWY	HKMQV	ILNOT	JMRSX	LMNPX	NQRTU
GNVWX	HKMSY	ILNSV	JNPRU	LMOPW	NQVWY
GOPRW	HKNPV	ILPRX	JNPTX	LMORU	NRSTX
GOPTU	HKNRY	ILUWY	JNUVW	LMOTX	NSUVY
GORTX	HKNTW	ILVWX	JNVXY	LMQUV	OPQTV
GOUVY	HKOTV	IMNOS	JOPRY	LMQWY	OPRSV
GPQVW	HKRST	IMOWX	JOPTW	LMSUY	OPSTY
GPSUV	HLMNR	IMPQX	JORTU	LMSVX	OQRTY
GPSWY	HLNTV	INUVX	JOVWY	LNPUV	ORSTW
GQRUV	HMNOT	INWXY	JPQVY	LNPWY	PQRST
GQRWY	HMNSV	IOPRU	JPSVW	LNRUY	PRUXY
GQTUY	HMPRX	IOPTX	JQRVW	LNRVX	PTUWX
GQTVX	HMUWY	IOUVW	JQTUV	LNTUW	QRSVY
GRSUY	HMVWX	IOVXY	JQTWY	LNTXY	QSTVW
GRSVX	HPRUV	IPQUV	JRSUV	LOPVY	UVWXY
GSTUW	HPRWY	IPQWY	JRSWY	LORVW	
GSTXY	HPTUY	IPSUY	JSTUY	LOTUV	
HIJPV	HPTVX	IPSVX	JSTVX	LOTWY	
HIJRY	HRTUW	IQRUY	KLMNO	LPQRV	

Total Combinations Of Five Squares That Add Up To A Selected Total
(Complex Formula)

The following **1,342** combinations of five squares add up to the magic total in a magic square whose "magic total" has a remainder of **one** once 65 has been subtracted from that magic total and the result of that subtraction has been divided by five. (Note that those combinations shown in bold print are those that are common to all 5 x 5 magic squares.)

ABCDE	**ACDJV**	ADNOT	**AFSVW**	AKMXY	BCKPS
ABCFW	**ACDKP**	**ADNUY**	AFTWX	**AKNPR**	**BCKXY**
ABCGV	**ACDLT**	ADPRX	AGHMT	**AKNVW**	**BCLMX**
ABCIY	**ACDMS**	**ADPSY**	**AGIUV**	**AKRSV**	BCLNY
ABCJM	**ACDNR**	**ADRSW**	**AGIWY**	AKRTX	BCLST
ABCKR	**ACEMR**	**ADSTU**	**AGJMW**	AKSTY	**BCMOU**
ABCNT	ACETV	ADVWX	**AGJPT**	**ALMNY**	BCNRS
ABDGX	**ACFGR**	**AEFGP**	**AGJVY**	**ALMST**	**BCNVX**
ABDHI	**ACFIP**	**AEFJR**	**AGKPY**	**ALNRT**	**BCOPR**
ABDJL	ACFJT	**AEFKW**	**AGKRW**	**AMNRS**	**BCOVW**
ABDNS	**ACFKY**	**AEFLV**	**AGKTU**	AMNVX	**BCPQV**
ABDOP	**ACFLM**	AEFMX	**AGLMP**	AMOPR	BCPUW
ABDQY	ACFNV	**AEFNY**	**AGLRV**	AMOVW	**BCQRY**
ABEFU	**ACGIT**	**AEFST**	**AGLTY**	AMPQV	**BCQTW**
ABEGY	**ACGKM**	**AEGIR**	AGMRX	**AMPUW**	**BDEHU**
ABEIW	**ACIKV**	**AEGKV**	**AGMSY**	AMQRY	**BDELQ**
ABEJV	**ACIMN**	**AEGMN**	**AGNPV**	AMQTW	**BDEOS**
ABEKP	**ACMUV**	**AEIJT**	**AGNRY**	**ANSTV**	**BDGHS**
ABELT	**ACMWY**	**AEIKY**	**AGNTW**	AOPTV	BDHKX
ABEMS	**ACPRV**	**AEILM**	**AGRST**	AORTY	**BDHLN**
ABENR	**ACPTY**	**AEINV**	AGTVX	**APUVY**	BDHOY
ABFGS	**ACRTW**	**AEJKM**	AHIMR	AQTVY	BDHQR
ABFKX	**ADEGW**	**AEKNT**	AHITV	**ARUVW**	**BDJQU**
ABFLN	**ADEIU**	AEMOT	AHKMV	**ATUWY**	**BDJSX**
ABFOY	**ADEJY**	**AEMUY**	**AIJMU**	**BCDHW**	BDKOQ
ABFQR	**ADELR**	**AEPRY**	**AIJPR**	**BCDJO**	**BDLOU**
ABGIJ	**ADENP**	**AEPTW**	**AIJVW**	**BCDNQ**	BDOWX
ABGKN	**ADESV**	**AERTU**	AIKMO	**BCDUX**	BDPQX
ABGMO	ADETX	**AEVWY**	**AIKPW**	**BCEFX**	**BDQSW**
ABGPW	ADFHK	**AFGIL**	**AIKRU**	BCEGN	**BEFJQ**
ABGRU	**ADFJS**	**AFGJK**	**AILPV**	BCEIL	**BEFLO**
ABHMY	ADFNX	AFGMQ	**AILRY**	BCEJK	**BEGHR**
ABHRT	ADFOW	AFGOT	**AILTW**	**BCEMQ**	**BEGIQ**
ABIKL	ADFPQ	**AFGUY**	AIMPX	**BCEOT**	BEGJU
ABIOV	**ADGIS**	AFHMP	**AIMSW**	BCEUY	**BEGKO**
ABIPU	**ADGKL**	AFHRV	**AINPY**	**BCFGQ**	**BEGPX**
ABIQT	ADGOV	AFHTY	**AINRW**	**BCFHP**	BEGSW
ABJPY	**ADGPU**	**AFIJN**	**AINTU**	**BCFNO**	**BEHIP**
ABJRW	ADGQT	**AFIKS**	**AIPST**	BCFSU	**BEHJT**
ABJTU	ADHMW	AFIOR	AIRVX	**BCGHT**	**BEHKY**
ABKMQ	ADHPT	AFIQV	**AISVY**	BCGJW	**BEHLM**
ABKOT	ADHVY	**AFIUW**	AITXY	BCGLP	**BEHNV**
ABKUY	ADIKX	**AFJUV**	**AJKPV**	**BCGRX**	**BEINO**
ABLMU	**ADILN**	**AFJWY**	**AJKRY**	BCGSY	BEISU
ABLPR	ADIOY	AFKOV	**AJKTW**	**BCHIR**	BEJLP
ABLVW	ADIQR	**AFKPU**	**AJLMR**	**BCHKV**	**BEJRX**
ABMWX	**ADJKN**	AFKQT	**AJLTV**	**BCHMN**	BEJSY
ABNUV	ADJMO	**AFLPY**	**AJMNP**	BCIJU	BEKLU
ABNWY	**ADJPW**	**AFLRW**	**AJMSV**	**BCIKO**	**BEKWX**
ABPSV	**ADJRU**	**AFLTU**	AJMTX	**BCIPX**	BELNW
ABPTX	ADKOR	AFMNO	**AJNRV**	BCISW	BELRS
ABRSY	ADKQV	**AFMSU**	**AJNTY**	BCJLR	**BELVX**
ABSTW	**ADKUW**	**AFNPW**	**AKLMW**	BCJNP	BENPS
ABVXY	**ADLUV**	**AFNRU**	**AKLPT**	BCJSV	**BENXY**
ACDFU	**ADLWY**	**AFPRS**	**AKLVY**	**BCJTX**	**BEOUV**
ACDGY	ADMNQ	AFPVX	**AKMNU**	BCKLW	**BEOWY**
ACDIW	ADMUX	AFRXY	**AKMPS**	BCKNU	**BEPQY**

Total Combinations Of Five Squares That Add Up To A Selected Total
(Complex Formula)

BEQRW	BIJOW	CDEKS	CFGHV	CIJNY	DEFQS
BEQTU	BIJPQ	CDEOR	CFGIO	CIJST	DEGHP
BESTX	BIKQU	CDEQV	CFGLW	CIKLP	DEGNO
BFGHL	BIKSX	CDEUW	CFGNU	CIKRX	DEGSU
BFHIX	BILNS	CDFJQ	CFGPS	CIKSY	DEHJR
BFHJN	BILOP	CDFLO	CFGXY	CILMS	DEHKW
BFHKS	BILQY	CDGHR	CFHIY	CILNR	DEHLV
BFHOR	BINQW	CDGIQ	CFHJM	CIMQW	DEHMX
BFHQV	BIORX	CDGJU	CFHKR	CINSV	DEHNY
BFHUW	BIOSY	CDGKO	CFHNT	CINTX	DEHST
BFIOQ	BIQRS	CDGPX	CFIKQ	CIOPV	DEIJQ
BFJLX	BIQVX	CDGSW	CFILU	CIORY	DEILO
BFJOU	BIUWX	CDHIP	CFIWX	CIOTW	DEJKO
BFLQW	BJKNS	CDHJT	CFJKU	CIPQT	DEJPX
BFNQU	BJKOP	CDHKY	CFJLY	CIQVY	DEJSW
BFNSX	BJKQY	CDHLM	CFJNW	CIUWY	DEKNQ
BFOPX	BJLMQ	CDHNV	CFJRS	CJKLT	DEKUX
BFOSW	BJLOT	CDINO	CFJVX	CJKMS	DELNU
BFPQS	BJLUY	CDISU	CFKPX	CJKNR	DELPS
BFQXY	BJMOS	CDJLP	CFKSW	CJMOR	DELXY
BGHIN	BJNOR	CDJRX	CFLNP	CJMQV	DEMOQ
BGHUV	BJNQV	CDJSY	CFLSV	CJMUW	DENWX
BGHWY	BJNUW	CDKLU	CFLTX	CJOTV	DEOUY
BGILX	BJPSW	CDKWX	CFMQU	CJPRW	DEPQW
BGIOU	BJRSU	CDLNW	CFMSX	CJPTU	DEQRU
BGJKX	BJUVX	CDLRS	CFNRX	CJUVY	DERSX
BGJLN	BJWXY	CDLVX	CFNSY	CKLNV	DFGHX
BGJOY	BKLOY	CDNPS	CFOPY	CKMNX	DFGOQ
BGJQR	BKLQR	CDNXY	CFORW	CKMOW	DFHJL
BGKLS	BKNOW	CDOUV	CFOTU	CKMPQ	DFHNS
BGKQW	BKNPQ	CDOWY	CFPQR	CKNST	DFHOP
BGLOR	BKORS	CDPQY	CFQVW	CKOPT	DFHQY
BGLQV	BKOVX	CDQRW	CGHIM	CKOVY	DFLQU
BGLUW	BKPUX	CDQTU	CGIJP	CKPUY	DFLSX
BGMQX	BKQSV	CDSTX	CGIKU	CKQRV	DFOSU
BGNOP	BKQTX	CEFGJ	CGILY	CKQTY	DFQWX
BGNQY	BKSUW	CEFIS	CGINW	CKRUW	DGHIL
BGOSV	BLNOV	CEFKL	CGIRS	CLMOV	DGHJK
BGOTX	BLNPU	CEFOV	CGIVX	CLMPU	DGHMQ
BGPSU	BLNQT	CEFPU	CGJKY	CLMQT	DGHOT
BGQST	BLPXY	CEFQT	CGJLM	CLPVW	DGHUY
BGUXY	BLRWX	CEGPY	CGJNV	CLRUV	DGJNX
BHIJK	BLSUV	CEGRW	CGKLR	CLRWY	DGJOW
BHIMQ	BLSWY	CEGTU	CGKNP	CLTUY	DGJPQ
BHIOT	BLTUX	CEHMV	CGKSV	CMNOY	DGKQU
BHIUY	BMNOX	CEIKN	CGKTX	CMNQR	DGKSX
BHJMU	BMNQS	CEIMO	CGLNT	CMOST	DGLNS
BHJPR	BMOPQ	CEIPW	CGMNS	CMPWX	DGLOP
BHJVW	BMSUX	CEIRU	CGMOP	CMRUX	DGLQY
BHKMO	BNOST	CEJPV	CGMQY	CMSUY	DGNQW
BHKPW	BNPWX	CEJRY	CGORV	CNORT	DGORX
BHKRU	BNRUX	CEJTW	CGOTY	CNPUV	DGOSY
BHLPV	BNSUY	CEKUV	CGPRU	CNPWY	DGQRS
BHLRY	BOPUY	CEKWY	CGQRT	CNQTV	DGQVX
BHLTW	BOQRV	CELMW	CGUVW	CNRUY	DGUWX
BHMPX	BOQTY	CELPT	CHIKT	CNTUW	DHIJN
BHMSW	BORUW	CELVY	CHMPY	CPRSY	DHIKS
BHNPY	BPQRU	CEMNU	CHMRW	CPSTW	DHIOR
BHNRW	BPRSX	CEMPS	CHMTU	CPVXY	DHIQV
BHNTU	BQUVW	CEMXY	CHPRT	CRSTU	DHIUW
BHPST	BSVWX	CENPR	CHRVY	CRVWX	DHJUV
BHRVX	CDEFH	CENVW	CHTVW	CSVWY	DHJWY
BHSVY	CDEGL	CERSV	CIJKW	CTUVX	DHKOV
BHTXY	CDEIX	CERTX	CIJLV	CTWXY	DHKPU
BIJNX	CDEJN	CESTY	CIJMX	DEFOX	DHKQT

Total Combinations Of Five Squares That Add Up To A Selected Total
(Complex Formula)

DHLPY	EFGLU	EIJVX	EPVWX	FJMQS	GILQT
DHLRW	EFGWX	EIKPX	ERUVX	FJNOY	GIMOX
DHLTU	EFHIW	EIKSW	ERWXY	FJNQR	GIMQS
DHMNO	EFHJV	EILNP	ESUVY	FJOST	GINOY
DHMSU	EFHKP	EILSV	ETUXY	FJPWX	GINQR
DHNPW	EFHLT	EILTX	FGHIJ	FJRUX	GIOST
DHNRU	EFHMS	EIMQU	FGHKN	FJSUY	GIPWX
DHPRS	EFHNR	EIMSX	FGHMO	FKLNS	GIRUX
DHPVX	EFIJO	EINRX	FGHPW	FKLOP	GISUY
DHRXY	EFINQ	EINSY	FGHRU	FKLQY	GJKOV
DHSVW	EFIUX	EIOPY	FGIQU	FKNQW	GJKPU
DHTWX	EFJLW	EIORW	FGISX	FKORX	GJKQT
DIJLX	EFJNU	EIOTU	FGJNS	FKOSY	GJLPY
DIJOU	EFJPS	EIPQR	FGJOP	FKQRS	GJLRW
DILQW	EFJXY	EIQVW	FGJQY	FKQVX	GJLTU
DINQU	EFKNO	EJKLR	FGKLX	FKUWX	GJMNO
DINSX	EFKSU	EJKNP	FGKOU	FLMOS	GJMSU
DIOPX	EFLRX	EJKSV	FGLOY	FLNOR	GJNPW
DIOSW	EFLSY	EJKTX	FGLQR	FLNQV	GJNRU
DIPQS	EFNPX	EJLNT	FGNOW	FLNUW	GJPRS
DIQXY	EFNSW	EJMNS	FGNPQ	FLPSW	GJPVX
DJKLS	EFOPW	EJMOP	FGORS	FLRSU	GJRXY
DJKQW	EFORU	EJMQY	FGOVX	FLUVX	GJSVW
DJLOR	EFQUV	EJORV	FGPUX	FLWXY	GJTWX
DJLQV	EFQWY	EJOTY	FGQSV	FMNQX	GKLMQ
DJLUW	EFSVX	EJPRU	FGQTX	FMOQW	GKLOT
DJMQX	EGHIV	EJQRT	FGSUW	FNOSV	GKLUY
DJNOP	EGHKT	EJUVW	FHIKL	FNOTX	GKMOS
DJNQY	EGILW	EKLMX	FHIOV	FNPSU	GKNOR
DJOSV	EGINU	EKLNY	FHIPU	FNQST	GKNQV
DJOTX	EGIPS	EKLST	FHIQT	FNUXY	GKNUW
DJPSU	EGIXY	EKMOU	FHJPY	FOPQT	GKPSW
DJQST	EGJKW	EKNRS	FHJRW	FOQVY	GKRSU
DJUXY	EGJLV	EKNVX	FHJTU	FOUWY	GKUVX
DKLNX	EGJMX	EKOPR	FHKMQ	FPQUY	GKWXY
DKLOW	EGJNY	EKOVW	FHKOT	FPSXY	GLMWX
DKLPQ	EGJST	EKPQV	FHKUY	FQRUW	GLNUV
DKNOU	EGKLP	EKPUW	FHLMU	FRSWX	GLNWY
DKOPS	EGKRX	EKQRY	FHLPR	FSTUX	GLPSV
DKOXY	EGKSY	EKQTW	FHLVW	GHIPY	GLPTX
DKQRX	EGLMS	ELMOY	FHMWX	GHIRW	GLRSY
DKQSY	EGLNR	ELMQR	FHNUV	GHITU	GLSTW
DLMOX	EGMQW	ELORT	FHNWY	GHJMP	GLVXY
DLMQS	EGNSV	ELPUV	FHPSV	GHJRV	GMNUX
DLNOY	EGNTX	ELPWY	FHPTX	GHJTY	GMOQV
DLNQR	EGOPV	ELQTV	FHRSY	GHKMU	GMOUW
DLOST	EGORY	ELRUY	FHSTW	GHKPR	GMPQU
DLPWX	EGOTW	ELTUW	FHVXY	GHKVW	GMPSX
DLRUX	EGPQT	EMNOW	FIJLS	GHLMY	GNPRX
DLSUY	EGQVY	EMNPQ	FIJQW	GHLRT	GNPSY
DNORS	EGUWY	EMORS	FILNX	GHMNW	GNRSW
DNOVX	EHIJM	EMOVX	FILOW	GHMRS	GNSTU
DNPUX	EHIKR	EMPUX	FILPQ	GHMVX	GNVWX
DNQSV	EHINT	EMQSV	FINOU	GHNPT	GOPRW
DNQTX	EHKMN	EMQTX	FIOPS	GHNVY	GOPTU
DNSUW	EHMPW	EMSUW	FIOXY	GHSTV	GOUVY
DOPQV	EHMRU	ENOPT	FIQRX	GIJKS	GPQVW
DOPUW	EHPVY	ENOVY	FIQSY	GIJOR	GQRUV
DOQRY	EHRVW	ENPUY	FJKNX	GIJQV	GQRWY
DOQTW	EHTUV	ENQRV	FJKOW	GIJUW	GQTUY
DQUWY	EHTWY	ENQTY	FJKPQ	GIKNX	GRSVX
DSUVX	EIJKU	ENRUW	FJLOV	GIKOW	GSTXY
DSWXY	EIJLY	EOSTV	FJLPU	GIKPQ	HIJPV
EFGHY	EIJNW	EPRSW	FJLQT	GILOV	HIJRY
EFGKQ	EIJRS	EPSTU	FJMOX	GILPU	HIJTW

Total Combinations Of Five Squares That Add Up To A Selected Total
(Complex Formula)

HIKUV	HMPQT	ILNUY	JLNVW	**KNRWX**	MOPXY
HIKWY	**HMQVY**	**ILPRX**	JLRSV	KNSUV	MORWX
HILMW	**HMUWY**	ILPSY	**JLRTX**	KNSWY	**MOSUV**
HILPT	HOTVY	ILRSW	JLSTY	**KNTUX**	**MOSWY**
HILVY	**HPRUV**	ILSTU	**JMNWX**	**KOPUV**	MOTUX
HIMNU	**HPRWY**	**ILVWX**	**JMOUY**	**KOPWY**	MPQRX
HIMPS	**HPTUY**	**IMNOS**	**JMPQW**	KOQTV	**MPQSY**
HIMXY	HQRTV	IMOQY	**JMQRU**	**KORUY**	**MQRSW**
HINPR	**HRTUW**	INPSW	**JMRSX**	**KOTUW**	**MQSTU**
HINVW	IJKLN	INRSU	JNPSV	**KPQRW**	**MQVWX**
HIRSV	**IJKOY**	**INUVX**	**JNPTX**	**KPQTU**	**NOPVW**
HIRTX	**IJKQR**	**INWXY**	JNRSY	**KPSTX**	**NORUV**
HISTY	**IJLMO**	**IOPRU**	JNSTW	**KQUVY**	**NORWY**
HJKMW	IJLPW	IOQRT	**JNVXY**	**KSVXY**	**NOTUY**
HJKPT	IJLRU	**IOUVW**	**JOPRY**	**LMNPX**	**NPQRY**
HJKVY	**IJNOV**	**IPQUV**	**JOPTW**	LMNSW	**NPQTW**
HJLMV	IJNPU	**IPQWY**	**JORTU**	**LMOPW**	**NQRTU**
HJMNY	**IJNQT**	**IPSVX**	**JOVWY**	**LMORU**	**NQVWY**
HJMST	**IJPXY**	**IQRUY**	**JPQVY**	**LMQUV**	**NRSTX**
HJNRT	**IJRWX**	**IQTUW**	JPUWY	**LMQWY**	**OPRSV**
HKLMP	IJSUV	**IRSXY**	**JQRVW**	**LMSVX**	OPRTX
HKLRV	IJSWY	**ISTWX**	**JQTUV**	LNPST	**OPSTY**
HKLTY	**IJTUX**	JKLUV	**JQTWY**	**LNRVX**	**ORSTW**
HKMRX	**IKLOR**	JKLWY	**JSTVX**	LNSVY	ORVXY
HKMSY	**IKLQV**	**JKMNQ**	**KLMNO**	**LNTXY**	OTVWX
HKNPV	IKLUW	**JKMUX**	KLMSU	**LOPVY**	**PQRST**
HKNRY	IKMQX	**JKNOT**	KLNPW	**LORVW**	PQTVX
HKNTW	**IKNOP**	JKNUY	KLNRU	**LOTUV**	**PRUXY**
HKRST	**IKNQY**	**JKPRX**	KLPRS	**LOTWY**	PSUVW
HKTVX	**IKOSV**	JKPSY	**KLPVX**	**LPQRV**	**PTUWX**
HLMNR	IKOTX	JKRSW	**KLRXY**	**LPQTY**	**QRSVY**
HLNTV	IKPSU	JKSTU	KLSVW	LPRUW	QRTXY
HMNSV	**IKQST**	**JKVWX**	**KLTWX**	**LQRTW**	**QSTVW**
HMNTX	**IKUXY**	JLMNU	KMOQR	**MNOPU**	RSUWY
HMOPV	**ILMNQ**	JLMPS	**KMQUW**	MNOQT	**UVWXY**
HMORY	**ILMUX**	**JLMXY**	**KMSWX**	**MNQUY**	
HMOTW	**ILNOT**	JLNPR	**KNPXY**	**MNSXY**	

Total Combinations Of Five Squares That Add Up To A Selected Total
(Complex Formula)

The following **1,291** combinations of five squares add up to the magic total in a magic square whose "magic total" has a remainder of **two** once 65 has been subtracted from that magic total and the result of that subtraction has been divided by five. (Note that those combinations shown in bold print are those that are common to all 5 x 5 magic squares.)

ABCDE	**ACDKP**	**ADPSY**	**AGJPT**	**AKSTY**	**BCQTW**
ABCFW	**ACDLT**	ADQVY	**AGJVY**	**ALMNY**	BCSUV
ABCGV	**ACDMS**	**ADRSW**	**AGKPY**	**ALMST**	BCSWY
ABCIY	**ACDNR**	**ADSTU**	**AGKRW**	**ALNRT**	**BDEHU**
ABCJM	**ACEMR**	**AEFGP**	**AGKTU**	AMNOT	**BDELQ**
ABCKR	**ACETV**	**AEFJR**	**AGLMP**	**AMNRS**	**BDEOS**
ABCNT	**ACFGR**	**AEFKW**	**AGLRV**	AMPRX	**BDGHS**
ABDHR	**ACFIP**	**AEFLV**	**AGLTY**	**AMPUW**	BDGOX
ABDIQ	**ACFJT**	**AEFNY**	AGMOV	AMVWX	BDHIO
ABDJL	**ACFKY**	**AEFST**	AGMQT	**ANSTV**	**BDHLN**
ABDKO	**ACFLM**	**AEGIR**	**AGMSY**	APTVX	BDHXY
ABDNS	**ACFNV**	**AEGKV**	**AGNPV**	**APUVY**	**BDJQU**
ABDPX	**ACGIT**	**AEGMN**	**AGNRY**	ARTXY	**BDJSX**
ABEFU	**ACGKM**	**AEIJT**	**AGNTW**	**ARUVW**	BDKQX
ABEGY	**ACIKV**	**AEIKY**	**AGRST**	**ATUWY**	**BDLOU**
ABEIW	**ACIMN**	**AEILM**	AHMPT	**BCDHW**	BDOQY
ABEJV	**ACMUV**	**AEINV**	AHMVY	**BCDJO**	**BDQSW**
ABEKP	**ACMWY**	**AEJKM**	AHRTV	**BCDNQ**	**BEFJQ**
ABELT	**ACPRV**	**AEKNT**	**AIJMU**	**BCDUX**	**BEFLO**
ABEMS	**ACPTY**	AEMTX	**AIJPR**	**BCEFX**	**BEGHR**
ABENR	**ACRTW**	**AEMUY**	**AIJVW**	BCEJP	**BEGIQ**
ABFGS	**ADEGW**	**AEPRY**	AIKMX	BCEKU	BEGJL
ABFHV	**ADEIU**	**AEPTW**	**AIKPW**	BCELY	**BEGKO**
ABFIO	**ADEJY**	**AERTU**	**AIKRU**	**BCEMQ**	BEGNS
ABFLN	**ADELR**	**AEVWY**	**AILPV**	BCENW	**BEGPX**
ABFXY	ADEMO	**AFGIL**	**AILRY**	**BCEOT**	**BEHIP**
ABGIJ	**ADENP**	**AFGJK**	**AILTW**	BCERS	**BEHJT**
ABGKN	**ADESV**	AFGTX	AIMOY	**BCFGQ**	**BEHKY**
ABGMX	ADFGO	**AFGUY**	AIMQR	**BCFHP**	**BEHLM**
ABGPW	ADFHY	AFHIT	**AIMSW**	BCFLS	**BEHNV**
ABGRU	**ADFJS**	AFHKM	**AINPY**	**BCFNO**	BEILS
ABHIM	ADFKQ	**AFIJN**	**AINRW**	**BCGHT**	**BEINO**
ABIKL	ADFWX	**AFIKS**	**AINTU**	BCGJN	BEJKS
ABIPU	ADGHM	AFIRX	AIORT	BCGKS	**BEJRX**
ABIVX	**ADGIS**	**AFIUW**	**AIPST**	**BCGRX**	BEJUW
ABJPY	**ADGKL**	**AFJUV**	AIQTV	BCGUW	**BEKWX**
ABJRW	**ADGPU**	**AFJWY**	**AISVY**	**BCHIR**	BELPU
ABJTU	ADGVX	**AFKPU**	**AJKPV**	**BCHKV**	**BELVX**
ABKTX	ADHIV	AFKVX	**AJKRY**	**BCHMN**	BENXY
ABKUY	ADHKT	**AFLPY**	**AJKTW**	BCIJL	**BEOUV**
ABLMU	**ADILN**	**AFLRW**	**AJLMR**	**BCIKO**	BEOWY
ABLPR	ADIXY	**AFLTU**	**AJLTV**	BCINS	BEPQY
ABLVW	**ADJKN**	AFMNX	**AJMNP**	**BCIPX**	**BEQRW**
ABMOP	ADJMX	AFMOW	**AJMSV**	**BCJTX**	BEQTU
ABMQY	**ADJPW**	AFMPQ	**AJNRV**	BCJUY	**BESTX**
ABNUV	**ADJRU**	**AFMSU**	**AJNTY**	BCKLN	BESUY
ABNWY	ADKRX	**AFNPW**	**AKLMW**	**BCKXY**	**BFGHL**
ABORV	**ADKUW**	**AFNRU**	**AKLPT**	**BCLMX**	BFHJN
ABOTY	**ADLUV**	AFOPT	**AKLVY**	BCLPW	BFHKS
ABPSV	**ADLWY**	AFOVY	**AKMNU**	BCLRU	BFHRX
ABQRT	ADMQW	**AFPRS**	AKMOR	**BCMOU**	**BFHUW**
ABRSY	ADNTX	AFQRV	**AKMPS**	BCNPU	BFIQX
ABSTW	**ADNUY**	AFQTY	AKMQV	**BCNVX**	BFJLX
ACDFU	ADOPV	**AFSVW**	**AKNPR**	**BCOPR**	BFJOU
ACDGY	ADORY	**AGIUV**	**AKNVW**	**BCOVW**	BFKOX
ACDIW	ADOTW	**AGIWY**	AKOTV	**BCPQV**	BFLQW
ACDJV	ADPQT	**AGJMW**	**AKRSV**	**BCQRY**	BFNQU

Total Combinations Of Five Squares That Add Up To A Selected Total
(Complex Formula)

BFNSX	BJMOS	CDKWX	CFOTU	CJRSV	DFHIQ
BFOQR	BJNOR	CDLPU	CFPQR	CJSTY	DFHJL
BFOSW	BJNPS	CDLVX	CFPSW	CKLPY	DFHKO
BFPQS	BJNQV	CDNXY	CFQVW	CKLRW	DFHNS
BGHIN	BJUVX	CDOUV	CFRSU	CKLTU	DFHPX
BGHUV	BJWXY	CDOWY	CGHIM	CKMNX	DFLQU
BGHWY	BKLOY	CDPQY	CGIKL	CKMOW	DFLSX
BGILX	BKLQR	CDQRW	CGIPU	CKMPQ	DFNOX
BGIOU	BKLSW	CDQTU	CGIVX	CKMSU	DFOPQ
BGJKX	BKMOQ	CDSTX	CGJPY	CKNPW	DFOSU
BGJOY	BKNOW	CDSUY	CGJRW	CKNRU	DGHIL
BGJQR	BKNPQ	CEFGU	CGJTU	CKOPT	DGHJK
BGJSW	BKNSU	CEFJW	CGKTX	CKOVY	DGHTX
BGKQW	BKORS	CEFLP	CGKUY	CKPRS	DGHUY
BGLNU	BKPUX	CEFOV	CGLMU	CKQRV	DGJNX
BGLOR	BKQSV	CEFQT	CGLPR	CKQTY	DGJOW
BGLPS	BLNOV	CEFSY	CGLVW	CKSVW	DGJPQ
BGLQV	BLNQT	CEGIW	CGMOP	CLMOV	DGJSU
BGNOP	BLNSY	CEGJV	CGMQY	CLMQT	DGKQU
BGNQY	BLPXY	CEGKP	CGNUV	CLMSY	DGKSX
BGOSV	BLRWX	CEGLT	CGNWY	CLNPV	DGLOP
BGQST	BLTUX	CEGMS	CGORV	CLNRY	DGLQY
BGUXY	BMNQS	CEGNR	CGOTY	CLNTW	DGNQW
BHIJK	BMOWX	CEHMV	CGPSV	CLRST	DGOQT
BHITX	BMPQX	CEIJY	CGQRT	CMNOY	DGOSY
BHIUY	BMSUX	CEILR	CGRSY	CMNQR	DGQRS
BHJMU	BNOST	CEIMO	CGSTW	CMNSW	DGUWX
BHJPR	BNPWX	CEINP	CHIKT	CMOST	DHIJN
BHJVW	BNRUX	CEISV	CHMPY	CMPWX	DHIKS
BHKMX	BOPTX	CEJKR	CHMRW	CMRUX	DHIRX
BHKPW	BOPUY	CEJNT	CHMTU	CNORT	DHIUW
BHKRU	BORUW	CEKLV	CHPRT	CNPST	DHJUV
BHLPV	BOVXY	CEKNY	CHRVY	CNQTV	DHJWY
BHLRY	BPQRU	CEKST	CHTVW	CNSVY	DHKPU
BHLTW	BPRSX	CELMN	CIJKN	CPRUW	DHKVX
BHMOY	BPSUW	CEMXY	CIJMX	CPVXY	DHLPY
BHMQR	BQRVX	CEPUV	CIJPW	CRVWX	DHLRW
BHMSW	BQTXY	CEPWY	CIJRU	CTUVX	DHLTU
BHNPY	BQUVW	CERTX	CIKRX	CTWXY	DHMNX
BHNRW	BSVWX	CERUY	CIKUW	DEFQS	DHMOW
BHNTU	CDEFH	CETUW	CILUV	DEGHP	DHMPQ
BHORT	CDEIX	CFGHV	CILWY	DEGLS	DHMSU
BHPST	CDELW	CFGIO	CIMQW	DEGNO	DHNPW
BHQTV	CDENU	CFGLN	CINTX	DEHJR	DHNRU
BHSVY	CDEOR	CFGXY	CINUY	DEHKW	DHOPT
BIJNX	CDEPS	CFHIY	CIOPV	DEHLV	DHOVY
BIJOW	CDEQV	CFHJM	CIORY	DEHNY	DHPRS
BIJPQ	CDFJQ	CFHKR	CIOTW	DEHST	DHQRV
BIJSU	CDFLO	CFHNT	CIPQT	DEIJQ	DHQTY
BIKQU	CDGHR	CFIJS	CIPSY	DEILO	DHSVW
BIKSX	CDGIQ	CFIKQ	CIQVY	DEJKO	DIJLX
BILOP	CDGJL	CFIWX	CIRSW	DEJNS	DIJOU
BILQY	CDGKO	CFJKL	CISTU	DEJPX	DIKOX
BINQW	CDGNS	CFJPU	CJKUV	DEKNQ	DILQW
BIOQT	CDGPX	CFJVX	CJKWY	DEKUX	DINQU
BIOSY	CDHIP	CFKNS	CJLMW	DELXY	DINSX
BIQRS	CDHJT	CFKPX	CJLPT	DEMQX	DIOQR
BIUWX	CDHKY	CFLTX	CJLVY	DENWX	DIOSW
BJKLU	CDHLM	CFLUY	CJMNU	DEOTX	DIPQS
BJKOP	CDHNV	CFMQU	CJMOR	DEOUY	DJKQW
BJKQY	CDILS	CFMSX	CJMPS	DEPQW	DJLNU
BJLMQ	CDINO	CFNRX	CJMQV	DEQRU	DJLOR
BJLNW	CDJKS	CFNUW	CJNPR	DERSX	DJLPS
BJLOT	CDJRX	CFOPY	CJNVW	DESUW	DJLQV
BJLRS	CDJUW	CFORW	CJOTV	DFGQX	DJNOP

Total Combinations Of Five Squares That Add Up To A Selected Total
(Complex Formula)

DJNQY	EGJMX	EKNPU	FHKUY	GHIPY	GMORX
DJOSV	EGJPW	EKNVX	FHLMU	GHIRW	GMOUW
DJQST	EGJRU	EKOPR	FHLPR	GHITU	GMPQU
DJUXY	EGKRX	EKOVW	FHLVW	GHJMP	GMPSX
DKLNX	EGKUW	EKPQV	FHMOP	GHJRV	GMQVX
DKLOW	EGLUV	EKQRY	FHMQY	GHJTY	GNPRX
DKLPQ	EGLWY	EKQTW	FHNUV	GHKMU	GNPUW
DKLSU	EGMQW	EKSUV	FHNWY	GHKPR	GNVWX
DKNOU	EGNTX	EKSWY	FHORV	GHKVW	GOPRW
DKOPS	EGNUY	ELMOY	FHOTY	GHLMY	GOPTU
DKOQV	EGOPV	ELMQR	FHPSV	GHLRT	GOTVX
DKQSY	EGORY	ELMSW	FHQRT	GHMNW	GOUVY
DLMQS	EGOTW	ELNPY	FHRSY	GHMOT	GPQVW
DLNOY	EGPQT	ELNRW	FHSTW	GHMRS	GPRSU
DLNQR	EGPSY	ELNTU	FIJQW	GHNPT	GQRUV
DLNSW	EGQVY	ELORT	FILNX	GHNVY	GQRWY
DLOST	EGRSW	ELPST	FILOW	GHSTV	GQTUY
DLPWX	EGSTU	ELQTV	FILPQ	GIJLW	GRSVX
DLRUX	EHIJM	ELSVY	FILSU	GIJNU	GSTXY
DMNOQ	EHIKR	EMNOW	FINOU	GIJOR	GSUVW
DMOUX	EHINT	EMNPQ	FIOPS	GIJPS	HIJPV
DNORS	EHKMN	EMNSU	FIOQV	GIJQV	HIJRY
DNPUX	EHMPW	EMORS	FIQSY	GIKNX	HIJTW
DNQSV	EHMRU	EMPUX	FJKNX	GIKOW	HIKUV
DOPRX	EHPVY	EMQSV	FJKOW	GIKPQ	HIKWY
DOPUW	EHRVW	ENOPT	FJKPQ	GIKSU	HILMW
DOVWX	EHTUV	ENOVY	FJKSU	GILOV	HILPT
DPQVX	EHTWY	ENPRS	FJLOV	GILQT	HILVY
DQRXY	EIJKL	ENQRV	FJLQT	GILSY	HIMNU
DQTWX	EIJPU	ENQTY	FJLSY	GIMQS	HIMOR
DQUWY	EIJVX	ENSVW	FJMQS	GINOY	HIMPS
DSUVX	EIKNS	EOSTV	FJNOY	GINQR	HIMQV
DSWXY	EIKPX	EPVWX	FJNQR	GINSW	HINPR
EFGHY	EILTX	ERUVX	FJNSW	GIOST	HINVW
EFGJS	EILUY	ERWXY	FJOST	GIPWX	HIOTV
EFGKQ	EIMQU	ETUXY	FJPWX	GIRUX	HIRSV
EFGWX	EIMSX	FGHIJ	FJRUX	GJKLP	HISTY
EFHIW	EINRX	FGHKN	FKLOP	GJKOV	HJKMW
EFHJV	EINUW	FGHMX	FKLQY	GJKQT	HJKPT
EFHKP	EIOPY	FGHPW	FKNQW	GJKSY	HJKVY
EFHLT	EIORW	FGHRU	FKOQT	GJLMS	HJLMV
EFHMS	EIOTU	FGIQU	FKOSY	GJLNR	HJMNY
EFHNR	EIPQR	FGISX	FKQRS	GJMNO	HJMST
EFIJO	EIPSW	FGJLU	FKUWX	GJNSV	HJNRT
EFINQ	EIQVW	FGJOP	FLMOS	GJPVX	HKLMP
EFIUX	EIRSU	FGJQY	FLNOR	GJRXY	HKLRV
EFJLN	EJKTX	FGKLX	FLNPS	GJTWX	HKLTY
EFJXY	EJKUY	FGKOU	FLNQV	GJUWY	HKMOV
EFKLS	EJLMU	FGLOY	FLUVX	GKLMQ	HKMQT
EFKNO	EJLPR	FGLQR	FLWXY	GKLNW	HKMSY
EFLRX	EJLVW	FGLSW	FMQWX	GKLOT	HKNPV
EFLUW	EJMOP	FGMOQ	FNOSV	GKLRS	HKNRY
EFMOX	EJMQY	FGNOW	FNQST	GKMOS	HKNTW
EFNPX	EJNUV	FGNPQ	FNUXY	GKNOR	HKRST
EFOPW	EJNWY	FGNSU	FOPVX	GKNPS	HLMNR
EFORU	EJORV	FGORS	FORXY	GKNQV	HLNTV
EFPSU	EJOTY	FGPUX	FOTWX	GKUVX	HMNSV
EFQUV	EJPSV	FGQSV	FOUWY	GKWXY	HMPVX
EFQWY	EJQRT	FHIKL	FPQTX	GLMWX	HMRXY
EFSVX	EJRSY	FHIPU	FPQUY	GLNST	HMTWX
EGHIV	EJSTW	FHIVX	FPSXY	GLPTX	HMUWY
EGHKT	EKLMX	FHJPY	FQRUW	GLPUY	HPRUV
EGILN	EKLPW	FHJRW	FQVXY	GLRUW	HPRWY
EGIXY	EKLRU	FHJTU	FRSWX	GLVXY	HPTUY
EGJKN	EKMOU	FHKTX	FSTUX	GMNUX	HRTUW

Total Combinations Of Five Squares That Add Up To A Selected Total
(Complex Formula)

HTVXY	**IMNOS**	JLPWY	**KLPVX**	**LMQWY**	**MQSTU**
IJKOY	IMOPX	**JLRTX**	**KLRXY**	**LMSVX**	**NOPVW**
IJKQR	IMQXY	JLRUY	**KLTWX**	LNPRU	**NORUV**
IJKSW	**INUVX**	JLTUW	KLUWY	**LNRVX**	**NORWY**
IJLMO	**INWXY**	**JMNWX**	KMOXY	**LNTXY**	**NOTUY**
IJLNP	**IOPRU**	JMOTX	**KMQRX**	LNUVW	**NPQRY**
IJLSV	IORVX	**JMOUY**	**KMQUW**	**LOPVY**	**NPQTW**
IJNOV	IOTXY	**JMPQW**	**KMSWX**	**LORVW**	NPSUV
IJNQT	**IOUVW**	**JMQRU**	**KNPXY**	**LOTUV**	NPSWY
IJNSY	**IPQUV**	**JMRSX**	**KNRWX**	**LOTWY**	**NQRTU**
IJPXY	**IPQWY**	JMSUW	**KNTUX**	**LPQRV**	**NQVWY**
IJRWX	**IPSVX**	**JNPTX**	**KOPUV**	**LPQTY**	**NRSTX**
IJTUX	IQRTX	JNPUY	**KOPWY**	LPSVW	NRSUY
IKLNU	**IQRUY**	JNRUW	KORTX	**LQRTW**	NSTUW
IKLOR	**IQTUW**	**JNVXY**	**KORUY**	LRSUV	**OPRSV**
IKLPS	**IRSXY**	**JOPRY**	**KOTUW**	LRSWY	**OPSTY**
IKLQV	**ISTWX**	**JOPTW**	**KPQRW**	LSTUY	OQTVY
IKNOP	ISUWY	**JORTU**	**KPQTU**	**MNOPU**	**ORSTW**
IKNQY	JKLNY	**JOVWY**	**KPSTX**	MNOVX	**PQRST**
IKOSV	JKLST	**JPQVY**	KPSUY	MNQTX	**PRUXY**
IKQST	**JKMNQ**	JPRSW	KQTVX	**MNQUY**	**PTUWX**
IKUXY	**JKMUX**	JPSTU	**KQUVY**	**MNSXY**	**QRSVY**
ILMNQ	**JKNOT**	**JQRVW**	KRSUW	MOPQV	**QSTVW**
ILMUX	JKNRS	**JQTUV**	**KSVXY**	MOQRY	**UVWXY**
ILNOT	**JKPRX**	**JQTWY**	**LMNPX**	MOQTW	
ILNRS	JKPUW	**JSTVX**	**LMOPW**	**MOSUV**	
ILPRX	**JKVWX**	JSUVY	**LMORU**	**MOSWY**	
ILPUW	**JLMXY**	**KLMNO**	LMPSU	**MPQSY**	
ILVWX	JLPUV	KLNSV	**LMQUV**	**MQRSW**	

Total Combinations Of Five Squares That Add Up To A Selected Total
(Complex Formula)

The following **1,243** combinations of five squares add up to the magic total in a magic square whose "magic total" has a remainder of **three** once 65 has been subtracted from that magic total and the result of that subtraction has been divided by five. (Note that those combinations shown in bold print are those that are common to all 5 x 5 magic squares.)

ABCDE	**ACDLT**	ADRXY	AGKRW	AMOTW	BDGHS
ABCFW	**ACDMS**	**ADSTU**	AGKTU	AMPQT	BDHIX
ABCGV	**ACDNR**	ADTWX	AGLMP	**AMPUW**	**BDHLN**
ABCIY	**ACEMR**	AEFGP	AGLRV	AMQVY	BDHOR
ABCJM	**ACETV**	AEFJR	AGLTY	**ANSTV**	BDHQV
ABCKR	**ACFGR**	AEFKW	AGMSY	AOTVY	**BDIOQ**
ABCNT	**ACFIP**	AEFLV	AGMVX	**APUVY**	**BDJQU**
ABDJL	**ACFJT**	AEFNY	AGNPV	AQRTV	**BDJSX**
ABDKX	**ACFKY**	AEFST	AGNRY	**ARUVW**	**BDLOU**
ABDNS	**ACFLM**	AEGIR	AGNTW	**ATUWY**	BDOPX
ABDOY	**ACFNV**	AEGKV	AGRST	**BCDHW**	**BDQSW**
ABDQR	**ACGIT**	AEGMN	AHIMV	**BCDJO**	BDQXY
ABEFU	**ACGKM**	**AEIJT**	AHKMT	**BCDNQ**	**BEFJQ**
ABEGY	**ACIKV**	AEIKY	AIJMU	**BCDUX**	**BEFLO**
ABEIW	**ACIMN**	AEILM	AIJPR	**BCEFX**	BEGHR
ABEJV	**ACMUV**	AEINV	AIJVW	BCEGJ	**BEGIQ**
ABEKP	**ACMWY**	AEJKM	AIKPW	BCEIS	**BEGKO**
ABELT	**ACPRV**	AEKNT	AIKRU	BCEKL	BEGLU
ABEMS	**ACPTY**	AEMUY	AILPV	**BCEMQ**	**BEGPX**
ABENR	**ACRTW**	AEPRY	AILRY	**BCEOT**	**BEHIP**
ABFGS	**ADEGW**	AEPTW	AILTW	BCEPU	**BEHJT**
ABFIX	**ADEIU**	AERTU	AIMSW	**BCFGQ**	BEHKY
ABFLN	**ADEJY**	AEVWY	AIMXY	**BCFHP**	**BEHLM**
ABFOR	**ADELR**	AFGIL	AINPY	**BCFNO**	BEHNV
ABFQV	ADEMX	AFGJK	AINRW	**BCGHT**	BEINO
ABGIJ	**ADENP**	AFGMO	AINTU	BCGLW	BEJLW
ABGKN	**ADESV**	AFGUY	AIPST	BCGNU	BEJNU
ABGPW	ADFGX	AFHMY	AIRTX	BCGPS	BEJPS
ABGRU	ADFHI	AFHRT	AISVY	**BCGRX**	**BEJRX**
ABHMR	**ADFJS**	**AFIJN**	AJKPV	BCHIR	BEKSU
ABHTV	ADFOP	**AFIKS**	AJKRY	BCHKV	**BEKWX**
ABIKL	ADFQY	AFIOV	AJKTW	BCHMN	BELSY
ABIMQ	**ADGIS**	AFIQT	AJLMR	BCIKO	**BELVX**
ABIOT	**ADGKL**	AFIUW	AJLTV	BCILU	BENSW
ABIPU	ADGMQ	AFJUV	AJMNP	**BCIPX**	BENXY
ABJPY	ADGOT	AFJWY	AJMSV	BCJKU	**BEOUV**
ABJRW	**ADGPU**	AFKMQ	AJNRV	BCJLY	**BEOWY**
ABJTU	ADHMP	AFKOT	AJNTY	BCJNW	BEPQY
ABKMO	ADHRV	**AFKPU**	AKLMW	BCJRS	BEQRW
ABKUY	**ADHTY**	AFLPY	AKLPT	**BCJTX**	BEQTU
ABLMU	**ADILN**	AFLRW	AKLVY	BCKSW	BESTX
ABLPR	ADIOR	AFLTU	AKMNU	**BCKXY**	BFGHL
ABLVW	ADIQV	AFMSU	AKMPS	**BCLMX**	BFHJN
ABMPX	**ADJKN**	AFMWX	AKMRX	BCLNP	BFHKS
ABNUV	**ADJPW**	AFNPW	AKNPR	BCLSV	BFHOV
ABNWY	**ADJRU**	AFNRU	AKNVW	**BCMOU**	BFHQT
ABPSV	ADKOV	AFPRS	AKRSV	BCNSY	**BFHUW**
ABRSY	ADKQT	AFPTX	AKSTY	**BCNVX**	BFJLX
ABRVX	**ADKUW**	AFSVW	AKTVX	**BCOPR**	BFJOU
ABSTW	**ADLUV**	AFVXY	ALMNY	**BCOVW**	BFLQW
ABTXY	**ADLWY**	AGIUV	ALMST	**BCPQV**	BFNQU
ACDFU	ADMNO	AGIWY	ALNRT	**BCQRY**	BFNSX
ACDGY	**ADNUY**	AGJMW	AMNRS	**BCQTW**	BFOSW
ACDIW	**ADPSY**	AGJPT	AMNTX	BDEHU	BFOXY
ACDJV	**ADPVX**	AGJVY	AMOPV	**BDELQ**	BFPQS
ACDKP	**ADRSW**	AGKPY	AMORY	BDEOS	BFQRX

Total Combinations Of Five Squares That Add Up To A Selected Total
(Complex Formula)

BGHIN	BKLQR	CEFGL	CGJMS	CLMUW	DGKSX
BGHUV	BKMQX	CEFJN	CGJNR	CLPRW	DGLOP
BGHWY	BKNOW	CEFKS	CGKLY	CLPTU	DGLQY
BGILX	BKNPQ	CEFOV	CGKNW	CLUVY	DGNQW
BGIOU	BKORS	CEFQT	CGKRS	CMNOY	DGOSY
BGJKX	BKOTX	CEFUW	CGKTX	CMNQR	DGOVX
BGJNS	BKPUX	CEGIN	CGLNV	CMOST	DGQRS
BGJOY	BKQSV	CEGUV	CGMOP	CMPSU	DGQTX
BGJQR	BLNOV	CEGWY	CGMQY	CMPWX	DGUWX
BGKQW	BLNQT	CEHMV	CGNST	CMRUX	DHIJN
BGLOR	BLNUW	CEIJK	CGORV	CNORT	DHIKS
BGLQV	BLPSW	CEIMO	CGOTY	CNPRU	DHIOV
BGMOX	BLPXY	CEIUY	CGPUY	CNQTV	DHIQT
BGNOP	BLRSU	CEJMU	CGQRT	CNUVW	DHIUW
BGNQY	BLRWX	CEJPR	CGRUW	CPSVW	DHJUV
BGOSV	BLTUX	CEJVW	CHIKT	CPVXY	DHJWY
BGQST	BMNQS	CEKPW	CHMPY	CRSUV	DHKMQ
BGSUW	BMOQY	CEKRU	CHMRW	CRSWY	DHKOT
BGUXY	BMSUX	CELPV	CHMTU	CRVWX	DHKPU
BHIJK	BNOST	CELRY	CHPRT	CSTUY	DHLPY
BHIMO	BNPSU	CELTW	CHRVY	CTUVX	DHLRW
BHIUY	BNPWX	CEMSW	CHTVW	CTWXY	DHLTU
BHJMU	BNRUX	CEMXY	CIJLR	DEFQS	DHMSU
BHJPR	BOPUY	CENPY	CIJMX	DEGHP	DHMWX
BHJVW	BOQRT	CENRW	CIJNP	DEGNO	DHNPW
BHKPW	BORUW	CENTU	CIJSV	DEHJR	DHNRU
BHKRU	BPQRU	CEPST	CIKLW	DEHKW	DHPRS
BHLPV	BPRSX	CERTX	CIKNU	DEHLV	DHPTX
BHLRY	BQUVW	CESVY	CIKPS	DEHNY	DHSVW
BHLTW	BSVWX	CFGHV	CIKRX	DEHST	DHVXY
BHMSW	CDEFH	CFGIO	CILNY	DEIJQ	DIJLX
BHMXY	CDEGS	CFGJU	CILST	DEILO	DIJOU
BHNPY	CDEIX	CFGSW	CIMQW	DEJKO	DILQW
BHNRW	CDELN	CFGXY	CINRS	DEJLU	DINQU
BHNTU	CDEOR	CFHIY	CINTX	DEJPX	DINSX
BHPST	CDEQV	CFHJM	CIOPV	DEKNQ	DIOSW
BHRTX	CDFJQ	CFHKR	CIORY	DEKUX	DIOXY
BHSVY	CDFLO	CFHNT	CIOTW	DELSW	DIPQS
BIJLS	CDGHR	CFIKQ	CIPQT	DELXY	DIQRX
BIJNX	CDGIQ	CFISU	CIPUW	DENSU	DJKQW
BIJOW	CDGKO	CFIWX	CIQVY	DENWX	DJLOR
BIJPQ	CDGLU	CFJLP	CJKLV	DEOUY	DJLQV
BIKQU	CDGPX	CFJSY	CJKNY	DEPQW	DJMOX
BIKSX	CDHIP	CFJVX	CJKST	DEQRU	DJNOP
BILOP	CDHJT	CFKLU	CJLMN	DERSX	DJNQY
BILQY	CDHKY	CFKPX	CJMOR	DFHJL	DJOSV
BINQW	CDHLM	CFLNW	CJMQV	DFHKX	DJQST
BIOSY	CDHNV	CFLRS	CJOTV	DFHNS	DJSUW
BIOVX	CDINO	CFLTX	CJPUV	DFHOY	DJUXY
BIQRS	CDJLW	CFMQU	CJPWY	DFHQR	DKLNX
BIQTX	CDJNU	CFMSX	CJRUY	DFKOQ	DKLOW
BIUWX	CDJPS	CFNPS	CJTUW	DFLQU	DKLPQ
BJKOP	CDJRX	CFNRX	CKLMS	DFLSX	DKNOU
BJKQY	CDKSU	CFOPY	CKLNR	DFOSU	DKOPS
BJLMQ	CDKWX	CFORW	CKMNX	DFOWX	DKORX
BJLOT	CDLSY	CFOTU	CKMOW	DFPQX	DKQSY
BJLPU	CDLVX	CFPQR	CKMPQ	DGHIL	DKQVX
BJMOS	CDNSW	CFQVW	CKNSV	DGHJK	DLMQS
BJNOR	CDNXY	CGHIM	CKOPT	DGHMO	DLNOY
BJNQV	CDOUV	CGIJW	CKOVY	DGHUY	DLNQR
BJSUY	CDOWY	CGILP	CKQRV	DGJLS	DLOST
BJUVX	CDPQY	CGISY	CKQTY	DGJNX	DLPSU
BJWXY	CDQRW	CGIVX	CKUWY	DGJOW	DLPWX
BKLNS	CDQTU	CGJKP	CLMOV	DGJPQ	DLRUX
BKLOY	CDSTX	CGJLT	CLMQT	DGKQU	DMNQX

Total Combinations Of Five Squares That Add Up To A Selected Total
(Complex Formula)

DMOQW	EHMRU	EPSUV	FJNOY	GIPSU	HISTY
DNORS	EHPVY	EPSWY	FJNQR	GIPWX	HITVX
DNOTX	EHRVW	EPVWX	FJOST	GIRUX	HJKMW
DNPUX	EHTUV	ERSUY	FJPSU	GJKOV	HJKPT
DNQSV	EHTWY	ERUVX	FJPWX	GJKQT	HJKVY
DOPQT	EIJLP	ERWXY	FJRUX	GJKUW	HJLMV
DOPUW	EIJSY	ESTUW	FKLOP	GJLUV	HJMNY
DOQVY	EIJVX	ETUXY	FKLQY	GJLWY	HJMST
DQUWY	EIKLU	FGHIJ	FKNQW	GJMNO	HJNRT
DSUVX	EIKPX	FGHKN	FKOSY	GJNUY	HKLMP
DSWXY	EILNW	FGHPW	FKOVX	GJPSY	HKLRV
EFGHY	EILRS	FGHRU	FKQRS	GJPVX	HKLTY
EFGKQ	EILTX	FGIQU	FKQTX	GJRSW	HKMSY
EFGSU	EIMQU	FGISX	FKUWX	GJRXY	HKMVX
EFGWX	EIMSX	FGJOP	FLMOS	GJSTU	HKNPV
EFHIW	EINPS	FGJQY	FLNOR	GJTWX	HKNRY
EFHJV	EINRX	FGKLX	FLNQV	GKLMQ	HKNTW
EFHKP	EIOPY	FGKOU	FLSUY	GKLOT	HKRST
EFHLT	EIORW	FGLNS	FLUVX	GKLPU	HLMNR
EFHMS	EIOTU	FGLOY	FLWXY	GKMOS	HLNTV
EFHNR	EIPQR	FGLQR	FMNOX	GKNOR	HMNSV
EFIJO	EIQVW	FGMQX	FMOPQ	GKNQV	HMOPT
EFINQ	EJKLY	FGNOW	FNOSV	GKSUY	HMOVY
EFIUX	EJKNW	FGNPQ	FNQST	GKUVX	HMQRV
EFJSW	EJKRS	FGORS	FNSUW	GKWXY	HMQTY
EFJXY	EJKTX	FGOTX	FNUXY	GLMSU	HMUWY
EFKNO	EJLNV	FGPUX	FOQRV	GLMWX	HORTV
EFLNU	EJMOP	FGQSV	FOQTY	GLNPW	HPRUV
EFLPS	EJMQY	FHIKL	FOUWY	GLNRU	HPRWY
EFLRX	EJNST	FHIMQ	FPQUY	GLPRS	HPTUY
EFNPX	EJORV	FHIOT	FPSXY	GLPTX	HRTUW
EFOPW	EJOTY	FHIPU	FQRUW	GLSVW	IJKNS
EFORU	EJPUY	FHJPY	FRSWX	GLVXY	IJKOY
EFQUV	EJQRT	FHJRW	FSTUX	GMNUX	IJKQR
EFQWY	EJRUW	FHJTU	GHIPY	GMOQT	IJLMO
EFSVX	EKLMX	FHKMO	GHIRW	GMOUW	IJLUY
EGHIV	EKLNP	FHKUY	GHITU	GMPQU	IJNOV
EGHKT	EKLSV	FHLMU	GHJMP	GMPSX	IJNQT
EGIJU	EKMOU	FHLPR	GHJRV	GNPRX	IJNUW
EGISW	EKNSY	FHLVW	GHJTY	GNSUV	IJPSW
EGIXY	EKNVX	FHMPX	GHKMU	GNSWY	IJPXY
EGJLR	EKOPR	FHNUV	GHKPR	GNVWX	IJRSU
EGJMX	EKOVW	FHNWY	GHKVW	GOPRW	IJRWX
EGJNP	EKPQV	FHPSV	GHLMY	GOPTU	IJTUX
EGJSV	EKQRY	FHRSY	GHLRT	GOUVY	IKLOR
EGKLW	EKQTW	FHRVX	GHMNW	GPQVW	IKLQV
EGKNU	ELMNS	FHSTW	GHMRS	GQRUV	IKMOX
EGKPS	ELMOY	FHTXY	GHMTX	GQRWY	IKNOP
EGKRX	ELMQR	FIJQW	GHNPT	GQTUY	IKNQY
EGLNY	ELORT	FILNX	GHNVY	GRSVX	IKOSV
EGLST	ELPRU	FILOW	GHSTV	GSTXY	IKQST
EGMQW	ELQTV	FILPQ	GIJLN	HIJPV	IKSUW
EGNRS	ELUVW	FINOU	GIJOR	HIJRY	IKUXY
EGNTX	EMNOW	FIOPS	GIJQV	HIJTW	ILMNQ
EGOPV	EMNPQ	FIORX	GIKLS	HIKUV	ILMUX
EGORY	EMORS	FIQSY	GIKNX	HIKWY	ILNOT
EGOTW	EMOTX	FIQVX	GIKOW	HILMW	ILNPU
EGPQT	EMPUX	FJKLS	GIKPQ	HILPT	ILPRX
EGPUW	EMQSV	FJKNX	GILOV	HILVY	ILSUV
EGQVY	ENOPT	FJKOW	GILQT	HIMNU	ILSWY
EHIJM	ENOVY	FJKPQ	GILUW	HIMPS	ILVWX
EHIKR	ENQRV	FJLOV	GIMQS	HIMRX	IMNOS
EHINT	ENQTY	FJLQT	GINOY	HINPR	IMOQR
EHKMN	ENUWY	FJLUW	GINQR	HINVW	INSUY
EHMPW	EOSTV	FJMQS	GIOST	HIRSV	INUVX

Total Combinations Of Five Squares That Add Up To A Selected Total
(Complex Formula)

INWXY	JLMXY	JQRVW	KORUY	LPQRV	NPQTW
IOPRU	JLNPY	JQTUV	KOTUW	LPQTY	NQRTU
IOQTV	JLNRW	JQTWY	KPQRW	LPUWY	NQVWY
IOUVW	JLNTU	JSTVX	KPQTU	LQRTW	NRSTX
IPQUV	JLPST	KLMNO	KPSTX	MNOPU	OPRSV
IPQWY	JLRTX	KLNUY	KQUVY	MNQUY	OPSTY
IPSVX	JLSVY	KLPSY	KSVXY	MNSXY	OPTVX
IQRUY	JMNSU	KLPVX	LMNPX	MOPRX	ORSTW
IQTUW	JMNWX	KLRSW	LMOPW	MOSUV	ORTXY
IRSXY	JMOUY	KLRXY	LMORU	MOSWY	PQRST
ISTWX	JMPQW	KLSTU	LMQUV	MOVWX	PRSUW
JKLPW	JMQRU	KLTWX	LMQWY	MPQSY	PRUXY
JKLRU	JMRSX	KMOQV	LMSVX	MPQVX	PTUWX
JKMNQ	JNPRS	KMQUW	LNPSV	MQRSW	QRSVY
JKMUX	JNPTX	KMSWX	LNRSY	MQRXY	QSTVW
JKNOT	JNSVW	KNPSW	LNRVX	MQSTU	QTVXY
JKNPU	JNVXY	KNPXY	LNSTW	MQTWX	UVWXY
JKPRX	JOPRY	KNRSU	LNTXY	NOPVW	
JKSUV	JOPTW	KNRWX	LOPVY	NORUV	
JKSWY	JORTU	KNTUX	LORVW	NORWY	
JKVWX	JOVWY	KOPUV	LOTUV	NOTUY	
JLMSW	JPQVY	KOPWY	LOTWY	NPQRY	

Total Combinations Of Five Squares That Add Up To A Selected Total
(Complex Formula)

The following **1,198** combinations of five squares add up to the magic total in a magic square whose "magic total" has a remainder of **four** once 65 has been subtracted from that magic total and the result of that subtraction has been divided by five. (Note that those combinations shown in bold print are those that are common to all 5 x 5 magic squares.)

ABCDE	**ACEMR**	**AEFKW**	**AGMSY**	ATVXY	**BEFLO**
ABCFW	**ACETV**	**AEFLV**	**AGNPV**	**BCDHW**	**BEGHR**
ABCGV	**ACFGR**	**AEFNY**	**AGNRY**	**BCDJO**	**BEGIQ**
ABCIY	**ACFIP**	**AEFST**	**AGNTW**	**BCDNQ**	BEGJS
ABCJM	**ACFJT**	**AEGIR**	**AGRST**	**BCDUX**	**BEGKO**
ABCKR	**ACFKY**	**AEGKV**	AHMRV	**BCEFX**	**BEGPX**
ABCNT	**ACFLM**	**AEGMN**	AHMTY	BCEGU	**BEHIP**
ABDHV	**ACFNV**	**AEIJT**	**AIJMU**	BCEJW	**BEHJT**
ABDIO	**ACGIT**	**AEIKY**	**AIJPR**	BCELP	**BEHKY**
ABDJL	**ACGKM**	**AEILM**	**AIJVW**	**BCEMQ**	**BEHLM**
ABDNS	**ACIKV**	**AEINV**	**AIKPW**	**BCEOT**	**BEHNV**
ABDXY	**ACIMN**	**AEJKM**	**AIKRU**	BCESY	**BEINO**
ABEFU	**ACMUV**	**AEKNT**	**AILPV**	**BCFGQ**	BEJLN
ABEGY	**ACMWY**	**AEMUY**	**AILRY**	**BCFHP**	**BEJRX**
ABEIW	**ACPRV**	**AEPRY**	**AILTW**	**BCFNO**	BEKLS
ABEJV	**ACPTY**	**AEPTW**	AIMOR	**BCGHT**	**BEKWX**
ABEKP	**ACRTW**	**AERTU**	AIMQV	BCGLN	BELUW
ABELT	**ADEGW**	**AEVWY**	**AIMSW**	**BCGRX**	**BELVX**
ABEMS	**ADEIU**	**AFGIL**	**AINPY**	**BCHIR**	**BENXY**
ABENR	**ADEJY**	**AFGJK**	**AINRW**	**BCHKV**	**BEOUV**
ABFGS	**ADELR**	AFGMX	**AINTU**	**BCHMN**	**BEOWY**
ABFHT	**ADENP**	**AFGUY**	AIOTV	BCIJS	**BEPQY**
ABFLN	**ADESV**	AFHIM	**AIPST**	**BCIKO**	BEPSU
ABFRX	ADFHR	**AFIJN**	AISVY	**BCIPX**	**BEQRW**
ABGIJ	ADFIQ	**AFIKS**	**AJKPV**	BCJKL	**BEQTU**
ABGKN	**ADFJS**	**AFIUW**	**AJKRY**	BCJPU	**BESTX**
ABGPW	ADFKO	AFIVX	**AJKTW**	**BCJTX**	**BFGHL**
ABGRU	ADFPX	**AFJUV**	**AJLMR**	BCKNS	**BFHJN**
ABIKL	**ADGIS**	**AFJWY**	**AJLTV**	**BCKXY**	**BFHKS**
ABIPU	**ADGKL**	**AFKPU**	**AJMNP**	**BCLMX**	**BFHUW**
ABITX	**ADGPU**	AFKTX	**AJMSV**	BCLUY	BFHVX
ABJPY	ADGTX	**AFLPY**	**AJNRV**	**BCMOU**	BFIOX
ABJRW	ADHIT	**AFLRW**	**AJNTY**	BCNUW	**BFJLX**
ABJTU	ADHKM	**AFLTU**	**AKLMW**	**BCNVX**	**BFJOU**
ABKMX	**ADILN**	AFMOP	**AKLPT**	**BCOPR**	**BFLQW**
ABKUY	ADIRX	AFMQY	**AKLVY**	**BCOVW**	**BFNQU**
ABLMU	**ADJKN**	**AFMSU**	**AKMNU**	**BCPQV**	**BFNSX**
ABLPR	**ADJPW**	**AFNPW**	AKMOV	BCPSW	BFOQV
ABLVW	**ADJRU**	**AFNRU**	**AKMPS**	**BCQRY**	**BFOSW**
ABMOY	**ADKUW**	AFORV	AKMQT	**BCQTW**	**BFPQS**
ABMQR	ADKVX	AFOTY	**AKNPR**	BCRSU	**BGHIN**
ABNUV	**ADLUV**	**AFPRS**	**AKNVW**	**BDEHU**	**BGHUV**
ABNWY	**ADLWY**	AFQRT	**AKRSV**	**BDELQ**	**BGHWY**
ABORT	ADMNX	**AFSVW**	**AKSTY**	**BDEOS**	**BGILX**
ABPSV	ADMOW	**AGIUV**	**ALMNY**	BDFHQ	**BGIOU**
ABQTV	ADMPQ	**AGIWY**	**ALMST**	**BDGHS**	**BGJKX**
ABRSY	**ADNUY**	**AGJMW**	**ALNRT**	BDHLN	BGJLU
ABSTW	ADOPT	**AGJPT**	**AMNRS**	BDHRX	**BGJOY**
ACDFU	ADOVY	**AGJVY**	**AMPUW**	BDIQX	**BGJQR**
ACDGY	**ADPSY**	**AGKPY**	AMPVX	**BDJQU**	**BGKQW**
ACDIW	ADQRV	**AGKRW**	AMRXY	**BDJSX**	**BGLOR**
ACDJV	ADQTY	**AGKTU**	AMTWX	BDKOX	**BGLQV**
ACDKP	**ADRSW**	**AGLMP**	**ANSTV**	**BDLOU**	BGLSW
ACDLT	**ADSTU**	**AGLRV**	**APUVY**	BDOQR	**BGNOP**
ACDMS	**AEFGP**	**AGLTY**	ARUVW	**BDQSW**	**BGNQY**
ACDNR	**AEFJR**	AGMOT	ATUWY	**BEFJQ**	BGNSU

Total Combinations Of Five Squares That Add Up To A Selected Total
(Complex Formula)

BGOSV	BNOST	CEKLR	CIJMX	CTWXY	DHNRU
BGQST	BNPWX	CEKNP	CIJUY	DEFQS	DHORV
BGUXY	BNRUX	CEKSV	CIKLN	DEGHP	DHOTY
BHIJK	BOPUY	CELNT	CIKRX	DEGNO	DHPRS
BHIMX	BORUW	CEMNS	CILPW	DEHJR	DHQRT
BHIUY	BORVX	CEMXY	CILRU	DEHKW	DHSVW
BHJMU	BOTXY	CEPRU	CIMQW	DEHLV	DIJLX
BHJPR	BPQRU	CERTX	CINPU	DEHNY	DIJOU
BHJVW	BPRSX	CEUVW	CINTX	DEHST	DILQW
BHKPW	BQRTX	CFGHV	CIOPV	DEIJQ	DINQU
BHKRU	BQUVW	CFGIO	CIORY	DEILO	DINSX
BHLPV	BSVWX	CFGJL	CIOTW	DEJKO	DIOQV
BHLRY	CDEFH	CFGNS	CIPQT	DEJPX	DIOSW
BHLTW	CDEIX	CFGXY	CIQVY	DEKNQ	DIPQS
BHMOR	CDEJU	CFHIY	CISUV	DEKUX	DJKQW
BHMQV	CDEOR	CFHJM	CISWY	DELNS	DJLOR
BHMSW	CDEQV	CFHKR	CJKPW	DELXY	DJLQV
BHNPY	CDESW	CFHNT	CJKRU	DEMOX	DJLSW
BHNRW	CDFJQ	CFIKQ	CJLPV	DENWX	DJNOP
BHNTU	CDFLO	CFILS	CJLRY	DEOUY	DJNQY
BHOTV	CDGHR	CFIWX	CJLTW	DEPQW	DJNSU
BHPST	CDGIQ	CFJKS	CJMOR	DEQRU	DJOSV
BHSVY	CDGJS	CFJUW	CJMQV	DERSX	DJQST
BIJNX	CDGKO	CFJVX	CJMSW	DFGOX	DJUXY
BIJOW	CDGPX	CFKPX	CJNPY	DFHIO	DKLNX
BIJPQ	CDHIP	CFLPU	CJNRW	DFHJL	DKLOW
BIKQU	CDHJT	CFLTX	CJNTU	DFHNS	DKLPQ
BIKSX	CDHKY	CFMQU	CJOTV	DFHXY	DKNOU
BILOP	CDHLM	CFMSX	CJPST	DFKQX	DKOPS
BILQY	CDHNV	CFNRX	CJSVY	DFLQU	DKOQT
BILSU	CDINO	CFOPY	CKLUV	DFLSX	DKQSY
BIMOQ	CDJLN	CFORW	CKLWY	DFOQY	DLMQS
BINQW	CDJRX	CFOTU	CKMNX	DFOSU	DLNOY
BIOSY	CDKLS	CFPQR	CKMOW	DGHIL	DLNQR
BIQRS	CDKWX	CFQVW	CKMPQ	DGHJK	DLOST
BIUWX	CDLUW	CFSUY	CKNUY	DGHMX	DLPWX
BJKOP	CDLVX	CGHIM	CKOPT	DGHUY	DLRUX
BJKQY	CDNXY	CGIJN	CKOVY	DGJNX	DMQWX
BJKSU	CDOUV	CGIKS	CKPSY	DGJOW	DNORS
BJLMQ	CDOWY	CGIUW	CKQRV	DGJPQ	DNPUX
BJLOT	CDPQY	CGIVX	CKQTY	DGKQU	DNQSV
BJLSY	CDPSU	CGJUV	CKRSW	DGKSX	DOPUW
BJMOS	CDQRW	CGJWY	CKSTU	DGLOP	DOPVX
BJNOR	CDQTU	CGKPU	CLMNU	DGLQY	DORXY
BJNQV	CDSTX	CGKTX	CLMOV	DGLSU	DOTWX
BJNSW	CEFLW	CGLPY	CLMPS	DGMOQ	DPQTX
BJUVX	CEFNU	CGLRW	CLMQT	DGNQW	DQUWY
BJWXY	CEFOV	CGLTU	CLNPR	DGOSY	DQVXY
BKLOY	CEFPS	CGMOP	CLNVW	DGQRS	DSUVX
BKLQR	CEFQT	CGMQY	CLRSV	DGUWX	DSWXY
BKNOW	CEGJR	CGMSU	CLSTY	DHIJN	EFGHY
BKNPQ	CEGKW	CGNPW	CMNOY	DHIKS	EFGKQ
BKORS	CEGLV	CGNRU	CMNQR	DHIUW	EFGLS
BKPUX	CEGNY	CGORV	CMOST	DHIVX	EFGWX
BKQSV	CEGST	CGOTY	CMPWX	DHJUV	EFHIW
BLNOV	CEHMV	CGPRS	CMRUX	DHJWY	EFHJV
BLNPS	CEIJP	CGQRT	CNORT	DHKPU	EFHKP
BLNQT	CEIKU	CGSVW	CNPSV	DHKTX	EFHLT
BLPXY	CEILY	CHIKT	CNQTV	DHLPY	EFHMS
BLRWX	CEIMO	CHMPY	CNRSY	DHLRW	EFHNR
BLTUX	CEINW	CHMRW	CNSTW	DHLTU	EFIJO
BMNQS	CEIRS	CHMTU	CPUWY	DHMOP	EFINQ
BMOPX	CEJKY	CHPRT	CPVXY	DHMQY	EFIUX
BMQXY	CEJLM	CHRVY	CRVWX	DHMSU	EFJNS
BMSUX	CEJNV	CHTVW	CTUVX	DHNPW	EFJXY

Total Combinations Of Five Squares That Add Up To A Selected Total
(Complex Formula)

EFKNO	EJORV	FHJRW	FSTUX	GMQTX	IJNQT
EFLRX	EJOTY	FHJTU	GHIPY	GNPRX	IJPXY
EFNPX	EJPRS	FHKMX	GHIRW	GNVWX	IJRWX
EFOPW	EJQRT	FHKUY	GHITU	GOPRW	IJTUX
EFORU	EJSVW	FHLMU	GHJMP	GOPTU	IKLOR
EFQUV	EKLMX	FHLPR	GHJRV	GOUVY	IKLQV
EFQWY	EKLUY	FHLVW	GHJTY	GPQVW	IKLSW
EFSUW	EKMOU	FHMOY	GHKMU	GPSUY	IKNOP
EFSVX	EKNUW	FHMQR	GHKPR	GQRUV	IKNQY
EGHIV	EKNVX	FHNUV	GHKVW	GQRWY	IKNSU
EGHKT	EKOPR	FHNWY	GHLMY	GQTUY	IKOSV
EGIJL	EKOVW	FHORT	GHLRT	GRSUW	IKQST
EGINS	EKPQV	FHPSV	GHMNW	GRSVX	IKUXY
EGIXY	EKPSW	FHQTV	GHMRS	GSTXY	ILMNQ
EGJMX	EKQRY	FHRSY	GHNPT	HIJPV	ILMUX
EGJUY	EKQTW	FHSTW	GHNVY	HIJRY	ILNOT
EGKLN	EKRSU	FIJQW	GHSTV	HIJTW	ILNSY
EGKRX	ELMOY	FILNX	GIJOR	HIKUV	ILPRX
EGLPW	ELMQR	FILOW	GIJQV	HIKWY	ILVWX
EGLRU	ELNUV	FILPQ	GIJSW	HILMW	IMNOS
EGMQW	ELNWY	FINOU	GIKNX	HILPT	IMOXY
EGNPU	ELORT	FIOPS	GIKOW	HILVY	IMQRX
EGNTX	ELPSV	FIOQT	GIKPQ	HIMNU	INUVX
EGOPV	ELQTV	FIQSY	GILNU	HIMOV	INWXY
EGORY	ELRSY	FJKNX	GILOV	HIMPS	IOPRU
EGOTW	ELSTW	FJKOW	GILPS	HIMQT	IORTX
EGPQT	EMNOW	FJKPQ	GILQT	HINPR	IOUVW
EGQVY	EMNPQ	FJLNU	GIMQS	HINVW	IPQUV
EGSUV	EMORS	FJLOV	GINOY	HIRSV	IPQWY
EGSWY	EMPUX	FJLPS	GINQR	HISTY	IPSUW
EHIJM	EMQSV	FJLQT	GIOST	HJKMW	IPSVX
EHIKR	ENOPT	FJMQS	GIPWX	HJKPT	IQRUY
EHINT	ENOVY	FJNOY	GIRUX	HJKVY	IQTUW
EHKMN	ENPSY	FJNQR	GJKLW	HJLMV	IQTVX
EHMPW	ENQRV	FJOST	GJKNU	HJMNY	IRSXY
EHMRU	ENQTY	FJPWX	GJKOV	HJMST	ISTWX
EHPVY	ENRSW	FJRUX	GJKPS	HJNRT	JKLNP
EHRVW	ENSTU	FKLOP	GJKQT	HKLMP	JKLSV
EHTUV	EOSTV	FKLQY	GJLNY	HKLRV	JKMNQ
EHTWY	EPVWX	FKLSU	GJLST	HKLTY	JKMUX
EIJKS	ERUVX	FKMOQ	GJMNO	HKMOT	JKNOT
EIJUW	ERWXY	FKNQW	GJNRS	HKMSY	JKNSY
EIJVX	ETUXY	FKOSY	GJPUW	HKNPV	JKPRX
EIKPX	FGHIJ	FKQRS	GJPVX	HKNRY	JKVWX
EILPU	FGHKN	FKUWX	GJRXY	HKNTW	JLMNS
EILTX	FGHPW	FLMOS	GJTWX	HKRST	JLMXY
EIMQU	FGHRU	FLNOR	GKLMQ	HLMNR	JLPRU
EIMSX	FGIQU	FLNQV	GKLOT	HLNTV	JLRTX
EINRX	FGISX	FLNSW	GKLSY	HMNSV	JLUVW
EIOPY	FGJOP	FLUVX	GKMOS	HMPTX	JMNWX
EIORW	FGJQY	FLWXY	GKNOR	HMUWY	JMOUY
EIOTU	FGJSU	FMOWX	GKNQV	HMVXY	JMPQW
EIPQR	FGKLX	FMPQX	GKNSW	HPRUV	JMQRU
EIQVW	FGKOU	FNOSV	GKUVX	HPRWY	JMRSX
EISUY	FGLOY	FNQST	GKWXY	HPTUY	JNPTX
EJKPU	FGLQR	FNUXY	GLMWX	HRTUW	JNUWY
EJKTX	FGNOW	FOPTX	GLNSV	HRTVX	JNVXY
EJLPY	FGNPQ	FOUWY	GLPTX	IJKLU	JOPRY
EJLRW	FGORS	FOVXY	GLUWY	IJKOY	JOPTW
EJLTU	FGPUX	FPQUY	GLVXY	IJKQR	JORTU
EJMOP	FGQSV	FPSXY	GMNUX	IJLMO	JOVWY
EJMQY	FHIKL	FQRUW	GMOUW	IJLNW	JPQVY
EJMSU	FHIPU	FQRVX	GMOVX	IJLRS	JPSUV
EJNPW	FHITX	FQTXY	GMPQU	IJNOV	JPSWY
EJNRU	FHJPY	FRSWX	GMPSX	IJNPS	JQRVW

Total Combinations Of Five Squares That Add Up To A Selected Total
(Complex Formula)

JQTUV	**KMSWX**	**KSVXY**	**LOTUV**	**MOSUV**	**NRSTX**
JQTWY	**KNPXY**	**LMNPX**	**LOTWY**	**MOSWY**	NSUVW
JRSUY	**KNRWX**	**LMOPW**	**LPQRV**	**MPQSY**	**OPRSV**
JSTUW	**KNTUX**	**LMORU**	**LPQTY**	**MQRSW**	**OPSTY**
JSTVX	**KOPUV**	**LMQUV**	LPRSW	**MQSTU**	OQRTV
KLMNO	**KOPWY**	**LMQWY**	LPSTU	**NOPVW**	**ORSTW**
KLNRS	**KORUY**	LMSUW	**LQRTW**	**NORUV**	**PQRST**
KLPUW	**KOTUW**	**LMSVX**	LSUVY	**NORWY**	**PRUXY**
KLPVX	KOTVX	LNPUY	**MNOPU**	**NOTUY**	**PTUWX**
KLRXY	**KPQRW**	LNRUW	MNOTX	**NPQRY**	**QRSVY**
KLTWX	**KPQTU**	**LNRVX**	**MNQUY**	**NPQTW**	**QSTVW**
KMORX	**KPSTX**	LNTXY	**MNSXY**	NPRSU	**UVWXY**
KMQUW	**KQUVY**	**LOPVY**	MOPQT	**NQRTU**	
KMQVX	KSUWY	**LORVW**	MOQVY	**NQVWY**	

Summary Of Number Of Combinations Of Five Squares That Add Up To A Selected Total (Simple Formula)

The following table shows the number of combinations of five squares that will sum to a specified "magic total" in a 5 x 5 magic square when the "simple" formula of creating the magic square is used:

"Magic Total"	Number Of Combinations
65	1,394
66	1,342
67	1,291
68	1,243
69	1,198
70	1,155
71	1,115
72	1,079
73	1,045
74	1,015
75	989
76	965
77	945
78	929
79	915
80	905
81	897
82	891
83	886
84	883
85	880
86	878
87	876
88	875
89	874
90	874
All Others	873

Computer Program To Generate An "Odd", "Basic" Magic Square

The following computer program, which I wrote and which is written in standard BASIC (Beginners' All Purpose Symbolic Instruction Code) - available on most personal computers - will generate any "odd", "basic" magic square, subject only to any limitations of computer memory and/or paper size (if the magic square is to be printed), using the formula detailed in Chapter 5, An Analysis Of 3 x 3 Magic Squares, in the above booklet:

```
10      INPUT X
20      IF (X<3) OR (INT(X/2)=X/2) THEN 10
30      DIM N(X*X)
40      FOR A=1 TO (X*X)
50      N(A)=0
60      NEXT A
70      Z=((X+1)/2)+(X-1)
80      FOR Y=1 TO (X*X)
90      Z1=Z-(INT(Z/X)*X)
100     IF Z1=0 THEN Z1=X
110     Z2=((Z-Z1)/X)+1
120     IF Z1=X THEN Z3=1 ELSE Z3=Z1+1
130     IF Z2=1 THEN Z4=X ELSE Z4=Z2-1
140     Z5=((Z4-1)*X)+Z3
150     IF N(Z5)<>0 THEN Z3=Z1:Z4=(Z2+1)-((Z2=X)*X)
160     Z5=((Z4-1)*X)+Z3
170     N(Z5)=Y
180     Z=Z5
190     NEXT Y
200     PRINT
210     PRINT "The following magic square adds up to ";((X*X*X)+X)/2
220     PRINT
230     FOR B=1 TO X*X
240     PRINT RIGHT$("      "+STR$(N(B)),5);
250     IF INT(B/X)=B/X THEN PRINT
260     NEXT B
```

Line 10 requests the user to enter a number, which represents the number of cells on each side of the magic square.

Line 20 checks the number that was entered and ensures that it is an odd number that is greater than two; if it fails this check, the program goes back to Line 10 to ask for another number to be entered.

Line 30 creates an empty magic square in memory with enough cells for the magic square that has been requested.

Computer Program To Generate An "Odd", "Basic" Magic Square

Lines 40 to 60 ensure that each cell in this empty magic square is set to an initial value of zero.

Line 70 calculates the cell that would contain the value zero, if such a value were present in the magic square; this is effectively the cell that would be before the cell that will contain the number one, and is derived as a start point for the following processing.

Line 80 is the start of an iterative piece of processing that will derive the value to be inserted in each of the cells of the magic square (i.e. x^2 cells).

Lines 90 to 160 calculate the cell that will contain the next value, using the formula explained in Chapter 5 of the above booklet.

Line 170 inserts the next value into the cell derived in Lines 90 to 160.

Line 180 is used to reset the value calculated in Line 70 so that the iterative processing may begin for the next cell and number.

Line 190 marks the end of the iterative processing, and will only stop once all x^2 cells have been filled with the numbers from one to x^2 inclusive, as appropriate.

Lines 200 - 220 prints a blank line, followed by a message stating that the magic square adds up to the "magic total" that is derived using the formula given above in Chapter 3, followed by another blank line.

Lines 230 - 260 loop round the generated magic square, printing "x" values per row, where "x" is the value entered in line 10 representing the number of cells on each side of the magic square.

Origami Magic Square Instructions

Here are instructions on how to fold the Crossed Box Pleat, using any square piece of paper:

1. With the writing or coloured side down (where appropriate), fold the square of paper in half, from top to bottom, unfold it, then fold it from left to right, and unfold that.

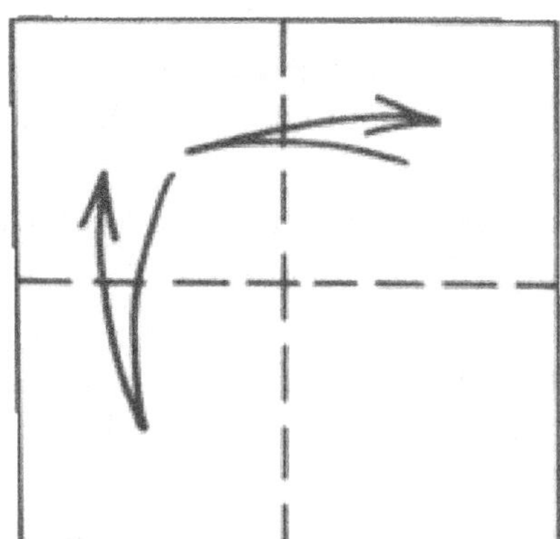

2. Still with the writing or coloured side down, fold the left edge to the centre crease, then unfold it, and then repeat by folding the right edge to the centre crease, and then unfold that.

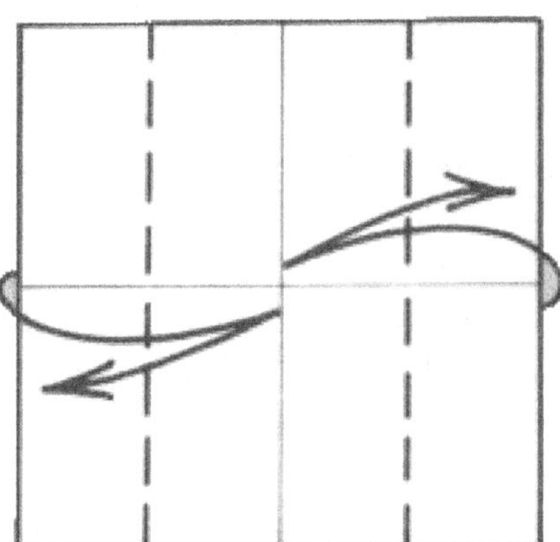

3. Still with the writing or coloured side down, fold the top edge to the centre crease, then unfold it, and then repeat by folding the bottom edge to the centre crease, and then unfold that.

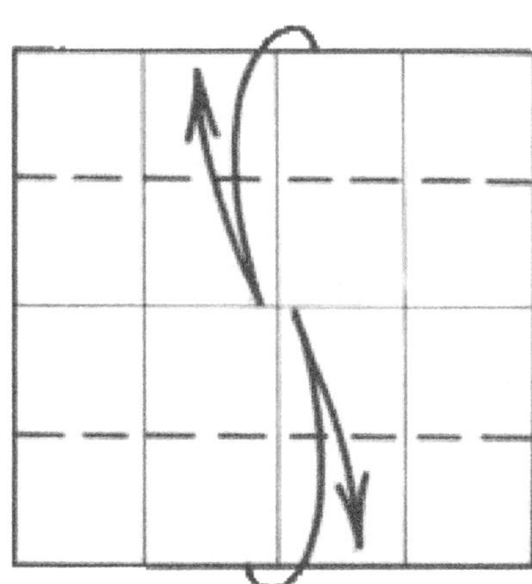

Origami Magic Square Instructions

4. Still with the writing or coloured side down, fold the right edge to the crease that is one quarter of the way from the left edge, then unfold it. Repeat this with the other three edges.

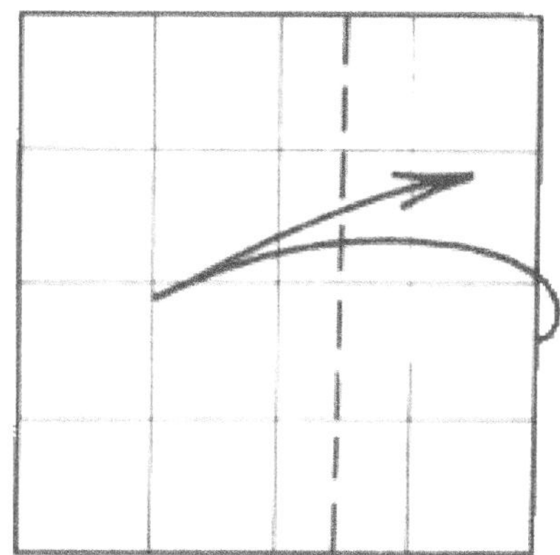

5. Turn the square of paper over and then fold the bottom right corner to the point that is three quarters of the way diagonally opposite, and then unfold it. Repeat this with the other four corners.

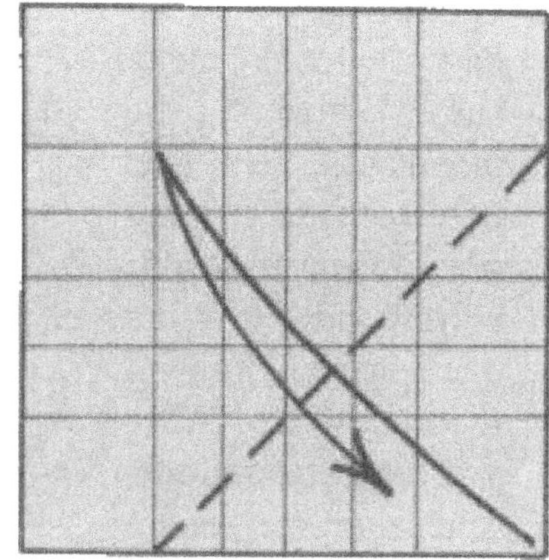

6. Fold the bottom right corner to the centre, and then unfold it. Repeat this with the other four corners.

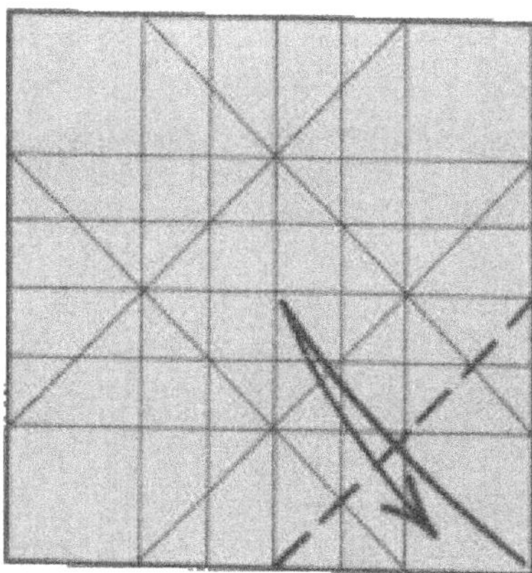

7. Open the square of paper out and flatten it.

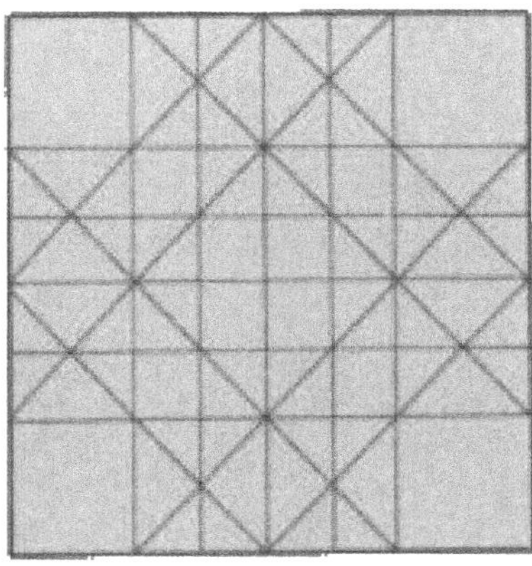

Origami Magic Square Instructions

8. No new folds are made in this step, but all of the existing folds need to be creased again sharply, as this will help in the next steps.

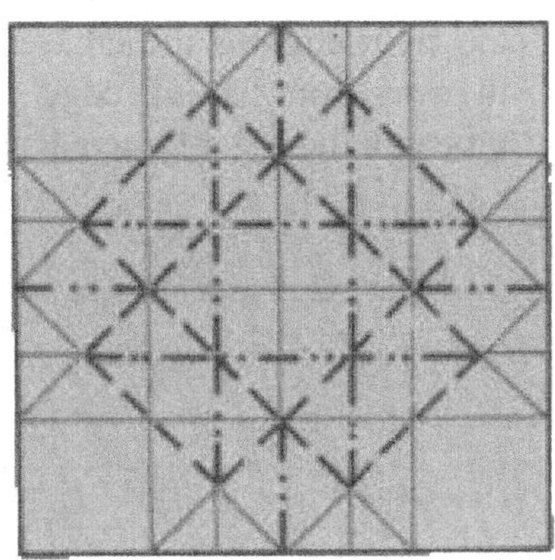

9. When pushing the edges together, make sure that the mountain (i.e. upward) and valley (i.e. downward folds correspond to the ones in the drawing to the right). After you have stressed this relief, turn the square of paper over.

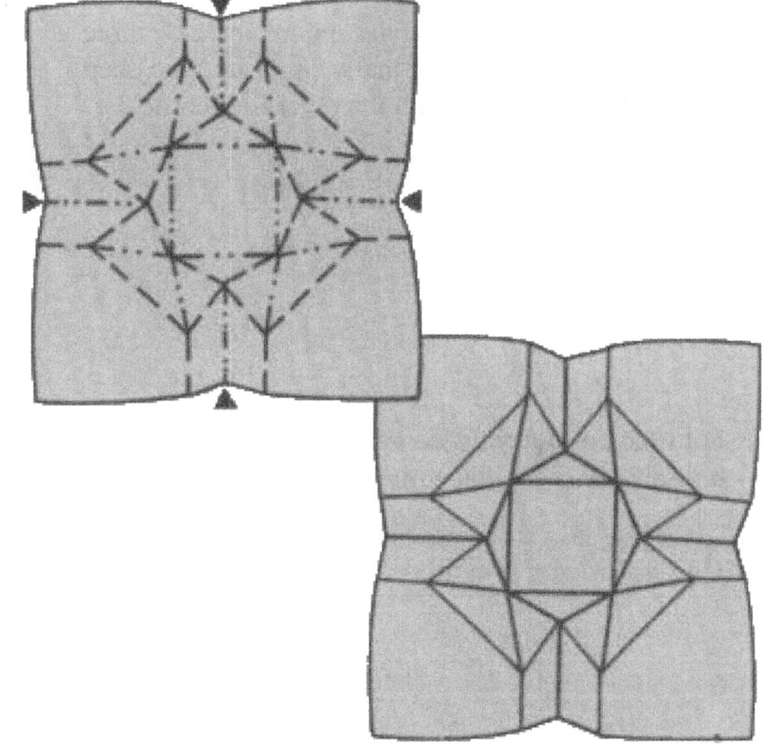

10. Push the parts together in the centre and then flatten the upstanding parts to the side.

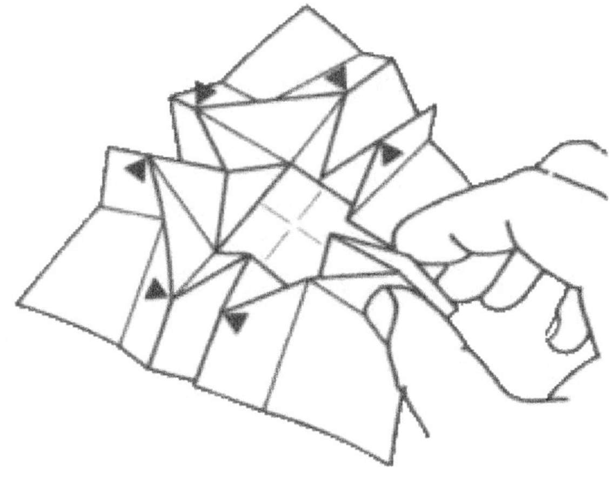

Origami Magic Square Instructions

11. This is the finished Crossed Box Pleat, seen from below.

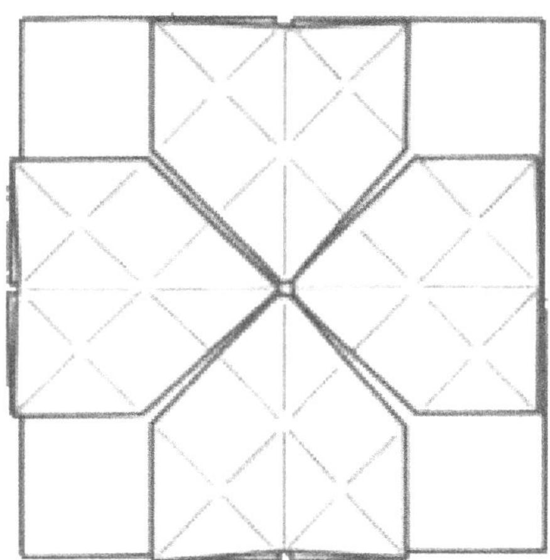

12. And this is the finished Crossed Box Pleat, seen from above.

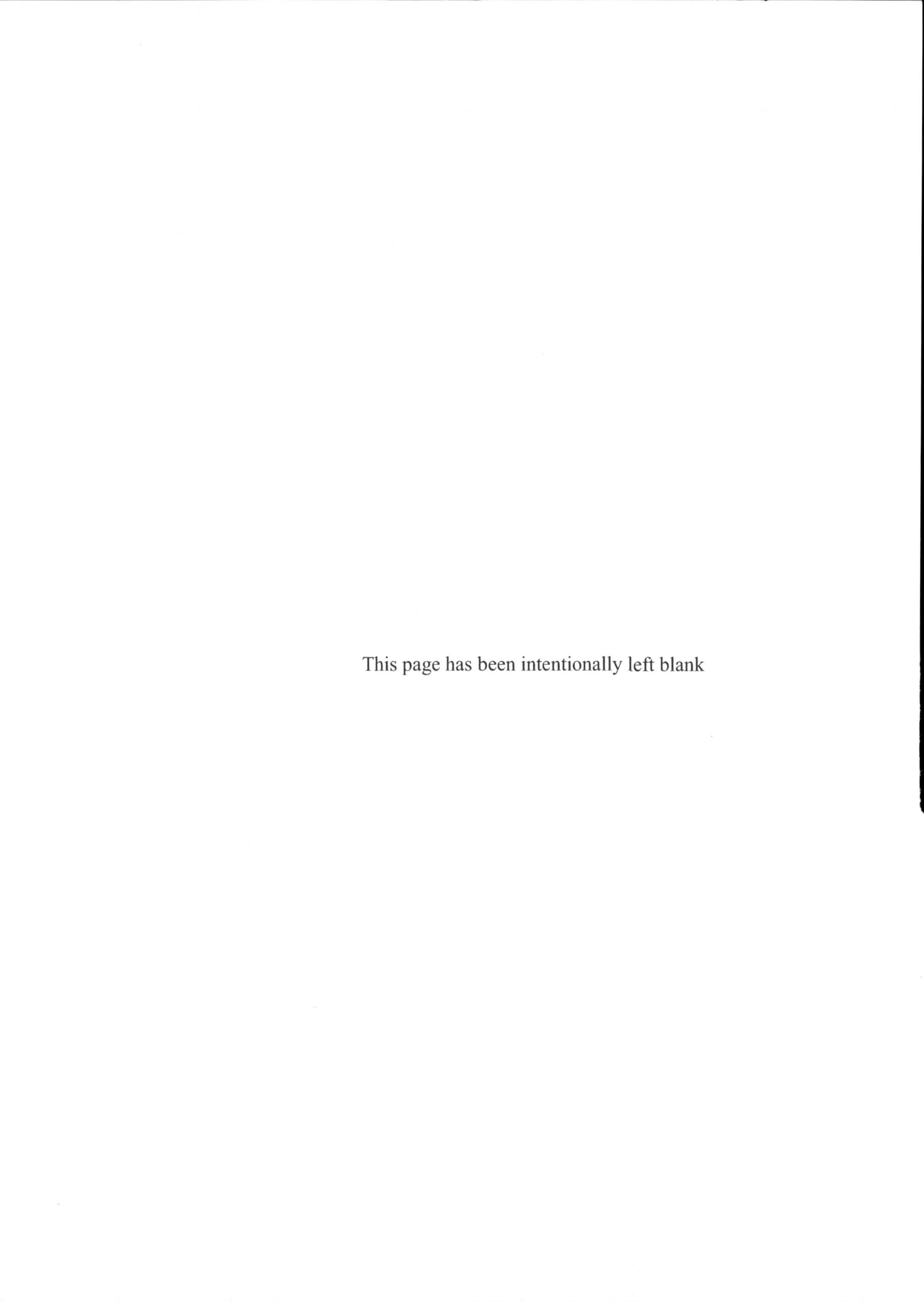

This page has been intentionally left blank

Origami Magic Square - Template

If you wish to perform "The Number Of The Beast", as described in Chapter 9, here is the template for the magic square, which you can cut out and photocopy, if you wish. (If you would rather not deface this book, then you can download the template from http://www.MagicSquaresBook.com/.)

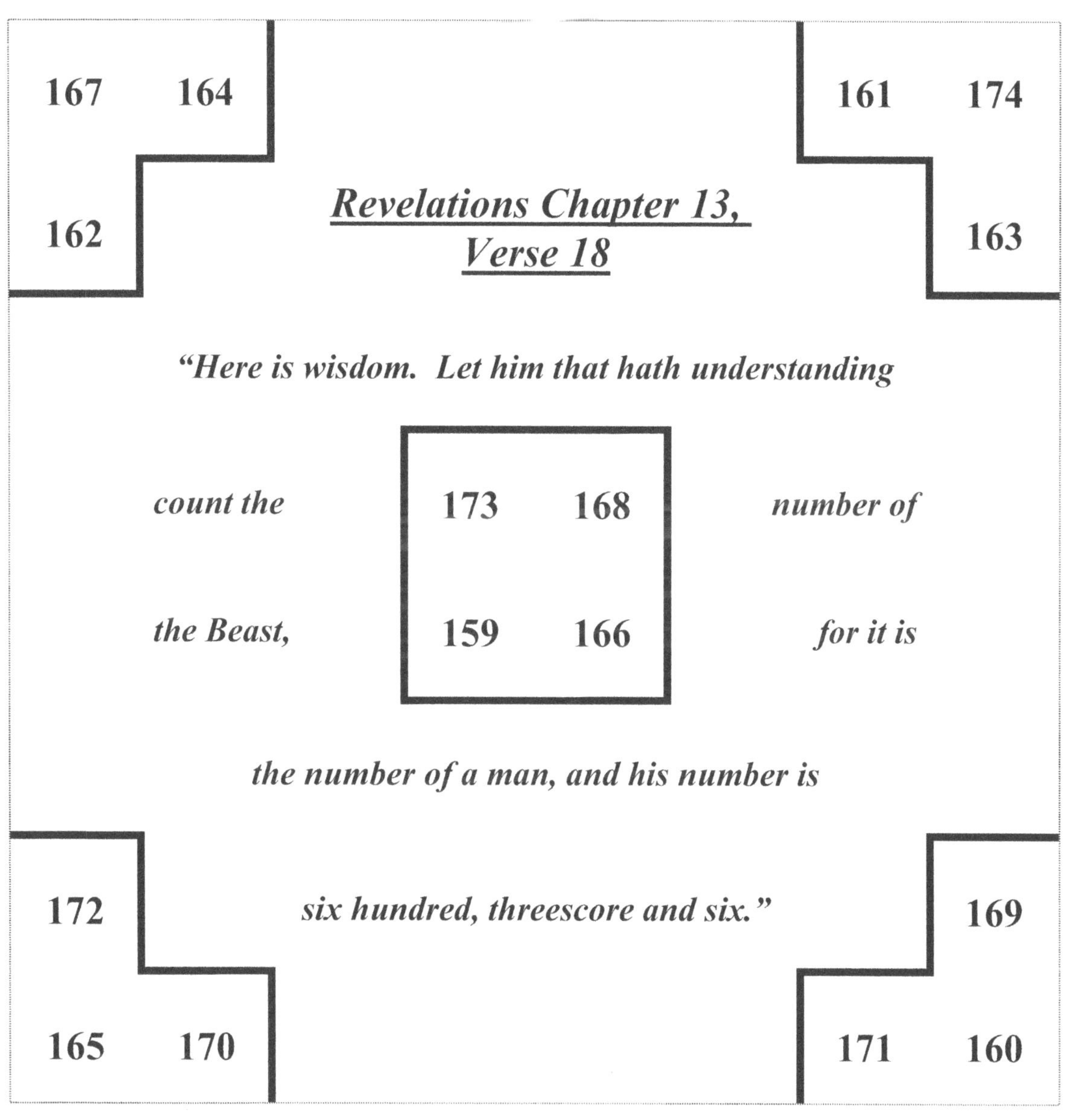

Additional Resources

Well, that's the end of the book, and I hope you have found it both informative and useful.

There are two additional resources that will be worth checking, from time to time:

1. http://www.MagicSquaresBook.com/

    This website will offer any updates and additions to this book, as well as allowing you to download the Origami Magic Square Template needed for "The Number Of The Beast" routine.

2. http://www.MagicSquares.org/

    This website provides a wide range of information on magic squares, including some that is in this book, as well as several interactive features.

Mark Farrar was born and bred in England, living there for the first 46 years of his life, before moving to the U.S.A. in 2005, where he now lives with his wife, Rae, and their family (of dogs).

Mark's background was in information technology, where he worked as a programmer, systems analyst and, lastly, business architect, and the technical skills developed during these years have served him well in the creation of many websites, including:

- www.MarkFarrar.co.uk
- www.4EverGifts4u.com
- www.LearningMagic.co.uk
- www.WealthyTeddy.co.uk

- www.AllKindsOfShiznit.com
- www.1StepMLM.com
- www.AromatherapyMadeEasy.com
- www.NicheMarketingCrashCourse.com

After receiving the usual boxed magic set as a young child, Mark's serious interest in the art of magic began in 1975, when he chanced upon one of England's leading unsung heroes of magic, the late Joe Riding.

It was also Joe that got Mark interested in magic squares, and it was this spark that finally led to the creation of this book, which was originally written as Mark's "entrance examination" to become a member of The Magic Circle, one of the oldest and most prestigious magic societies in the world, and one of the most well known. The book was so well received that Mark was not only admitted as a Member of The Magic Circle, but was also promoted to Associate Of The Inner Magic Circle on the strength of this work, something that very few members have achieved (most earning this recognition through performing ability rather than through more scholarly works).

Since then, copies of an earlier version of this book have been bought by magicians and mathematicians all over the world, and the library of The Magic Castle, in Los Angeles, specifically requested a copy to add to their own, massive collection.

Mark has also presented lectures on magic squares to both magical and non-magical audiences.

Mark has been involved in various magic societies during this period, with particular emphasis on the Northamptonshire Magicians' Club (www.N-M-C.org), where he was honoured by being made a Life Vice President in 2005, and The Magic Circle (www.TheMagicCircle.co.uk).

Made in the USA
Monee, IL
07 July 2026